ACS SYMPOSIUM SERIES **464**

Cell Separation Science and Technology

Dhinakar S. Kompala, EDITOR
University of Colorado

Paul Todd, EDITOR
National Institute of Standards and Technology

Developed from a symposium sponsored
by the Divisions of Industrial and Engineering Chemistry, Inc.,
and Biochemical Technology
at the 199th National Meeting
of the American Chemical Society,
Boston, Massachusetts,
April 22–27, 1990

American Chemical Society, Washington, DC 1991

Library of Congress Cataloging-in-Publication Data

Cell separation science and technology / Dhinakar S. Kompala, editor,
Paul Todd, editor.

    p.   cm.—(ACS symposium series; 464.)

"Developed from a symposium sponsored by the Divisions of
Industrial and Engineering Chemistry, Inc., and Biochemical Technology
at the 199th National Meeting of the American Chemical Society,
Boston, Massachusetts, April 22–27, 1990."

Includes bibliographical references and index.

ISBN 0–8412–2090–5

1. Cell separation—Congresses.

I. Kompala, Dhinakar S., 1958–   . II. Todd, Paul, 1936–   .
III. American Chemical Society. Division of Industrial and Engineering
Chemistry, Inc. IV. American Chemical Society. Division of
Biochemical Technology. V. American Chemical Society. Meeting
(199th: 1990: Boston, Mass.) VI. Series.

QH585.5.C44C45 1991   91–17033
574.87—dc20   CIP

ACS Symposium Series

M. Joan Comstock, *Series Editor*

1991 ACS Books Advisory Board

Foreword

THE ACS SYMPOSIUM SERIES was founded in 1974 to provide a medium for publishing symposia quickly in book form. The format of the Series parallels that of the continuing ADVANCES IN CHEMISTRY SERIES except that, in order to save time, the papers are not typeset, but are reproduced as they are submitted by the authors in camera-ready form. Papers are reviewed under the supervision of the editors with the assistance of the Advisory Board and are selected to maintain the integrity of the symposia. Both reviews and reports of research are acceptable, because symposia may embrace both types of presentation. However, verbatim reproductions of previously published papers are not accepted.

Contents

INDEXES

Preface

LARGE POPULATIONS OF SEPARATED CELLS are needed for many applications in biotechnology and biomedicine, including biochemical study, product analysis in nonclonogenic cells, selection of fused cells, isolation of rare cell types for cloning, preparation of pure cells for transplantation, and bioreactor maintenance. Any cell type that has limited or no proliferative capabilities but is needed in pure form must be separated by a technique that provides adequate purity, adequate yield, adequate relationship to cell function, and adequate function after separation.

Optical sorting, inertial methods (sedimentation and field-flow fractionation), affinity-based methods (adhesion, extraction, field-flow, and magnetic), and electrophoresis are the four principal modern methods of cell separation. Approximately 99 percent of all cell separations are performed in a centrifuge as a part of a research experiment. Other methods, such as magnetic separation, affinity adsorption (*panning*), electrophoresis, and single-cell selection, have existed for at least 50 years. However, only recently has the demand for large populations of separated cells in biotechnology and biomedical applications needed the development and application of a wider variety of methods. Biotechnology and biomedicine have now become significant activities in industrial and engineering chemistry. This book covers nearly all of the methods now used in the separation of living cells.

Acknowledgments

We acknowledge with deep gratitude the enthusiastic cooperation of all the authors and of Cheryl Shanks and the staff of the American Chemical Society Books Department for making this volume a reality.

DHINAKAR S. KOMPALA
University of Colorado
Boulder, CO 80309–0424

PAUL TODD
National Institute of Standards and Technology
Boulder, CO 80303–3328

March 15, 1991

Chapter 1

Separation of Living Cells

Paul Todd[1] and Thomas G. Pretlow[2]

[1]Center for Chemical Technology, National Institute of Standards and Technology, 325 Broadway 831.02, Boulder, CO 80303–3328
[2]Department of Pathology, Case Western Reserve University, Cleveland, OH 44106

There are several reasons for isolating subpopulations of living cells, including biochemical study, product analysis in non-clonogenic cells, selection of fused cells, isolation of rare cell types for cloning, preparation of pure cells for transplantation, and bioreactor maintenance. Methods for separating cells include flow sorting, sedimentation, biphasic extraction, field-flow fractionation, affinity methods, magnetically enhanced affinity methods, and electrophoresis. Each method can be considered in terms of its resolution, purity, gentleness, convenience and cost. Recent advances in each of these methods improves its utility in routine cell-separation practice and its potential for scale-up.

The chapters that follow cover, in one or more forms, nearly all of the methods used in the separation of living cells. This chapter introduces the rationales for separating living cells, the parameters with which separation methods are evaluated, and some physical characteristics of each of the methods reported in the subsequent chapters. Books containing review articles (1-4) appear from time to time on this subject. To date, most applications of living cell separation have occurred in the biomedical research field and in the field of cell and molecular biology. Applications to bioprocessing are still emerging.

Why separate living cells?

Novel products of biotechnology include particulate materials. These take the form of subcellular particles, bacterial inclusion bodies, whole cells, and insoluble macromolecular aggregates. Theoretically,

0097–6156/91/0464–0001$07.00/0

the logical way to purify these products is by centrifugation. In an
increasing number of cases of subcellular particles this approach
fails. Some very small particles have very low sedimentation rates
and hence require ultracentrifugation -- a satisfactory method only
at bench scale. Some particles are contaminated with unwanted
particles that have identical sedimentation coefficients. Processing
methods that can deal with these troublesome separation problems
include two-phase extraction, free electrophoresis, affinity
adsorption and field-flow fractionation. The reasons for purifying
suspended animal and plant cells are numerous, and the activity
itself is amply justified by the demands for pure cell populations as
objects of chemical research, living material for transplantation,
and sources of uncontaminated bioproducts. Although histochemical
studies are capable of demonstrating which antigens or which enzyme
activities are made by certain cells, they show very little
information about processing, molecular weight forms, inhibitors or
growth factors. Any cell type with limited or no proliferative
capabilities that is needed in pure form must be separated by a
technique that provides adequate purity, adequate yield, adequate
relationship to cell function, and adequate function after separation
(Todd et al., 1986). Light-activated flow sorting, which may process
a few hundred initial cells per second, is incapable of producing
adequate numbers of rare cell types for biochemical analysis. Cell
electrophoresis ($\underline{6}$), high gradient magnetic filtration ($\underline{7}$), two-phase
partitioning ($\underline{8}$, $\underline{9}$), and affinity adsorption ($\underline{10}$) have been
identified as suitable processes for cell separation. These
processes, although not especially new ($\underline{11}$, $\underline{12}$), are in the
adolescent phase of development, and processes yet to be discovered
may be applicable to the particle purification problem.

Relating Cell Structure and Function.   Size and density are
distributed values in any living cell population. (See Figure 1).
These may or may not be related to cell function. For example, mast
cells and somatotrophs are exceptionally dense owing to their densely
packed granules of secretory materials. Sedimentation helps separate
such cells from others with which they are naturally mixed.
Heterogeneity of sedimentation in a pure subpopulation interferes
with purity and yield in most purification processes, including
certain types of cell electrophoresis ($\underline{13}$). In density gradient
electrophoresis it has been shown that human embryonic kidney cells
migrate independently of size, density, and position in the cell
cycle ($\underline{5}$); and adult mammalian kidney cells can be separated by
electrophoresis into subpopulations that vary with respect to renin
production and other functions ($\underline{14}$, $\underline{15}$). Two-phase extraction, which
depends on cells partitioning into upper and lower immiscible aqueous
solutions of polymers, has been applied to cell separation problems
in hematology and immunology ($\underline{4}$).
   While histochemists routinely relate cell structure and function,
histochemistry alone can not reveal the subtle (and sometimes
substantial) differences that have been discovered by cell
purification, such as immunologically similar molecules with
different biological function produced by separable, morphologically

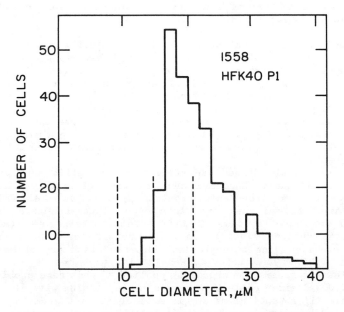

Figure 1. Size heterogeneity of human kidney cells in first-passage culture, measured microscopically. Dashed vertical lines indicate calibration points. The range of diameters is nearly a factor of 3.

similar, cells (16, 17). Histochemistry cannot measure the molecular weights of cell products or determine gene sequences in such cells; cell purification has made it possible to do these.   Owing to electrophoretic purification, we now know, for example, that two pituitary cell types make two types of growth hormone and that different electrophoretic subpopulations of kidney cells make plasminogen activator molecules having different properties.

The findings required procedures for reliably isolating several specific cell types from mammalian solid tissues (13,17).   The accomplishment of methodological goals such as these provides new means to highly significant ends: a way to study the functions of specific cells of tissues and effective methods of bulk cell separation.   The pharmacological study of separated cell populations also has the potential of reducing the amount of costly (and, to some, objectionable) animal research.

Clonogenic but Rare Plant and Animal Cells.   Most clonogenic cells from all living kingdoms can be isolated by classical methods consisting of selecting a whole colony of cells having the desired properties and propagating cells from it in continuous or batch culture.   Such techniques as replica plating, colony "picking" (loop, toothpick, pipette, selection ring, etc.), and dilution in microwells are used to isolate whole colonies that are presumed to have descended from a single cell. Selecting one cell with a micropipette under a microscope   also exploits the clonogenic potential of a desired cell type.

These classical methods are still the most efficient for most genetic cloning goals; however, when the desired cell type is less than 0.01% of the clonogenic population, then more than 10,000 colonies must be screened to obtain each clone unless a genetic selection technique is used.   Genetic selection techniques (killing all but the desired clones) can be contrived for most microbial cells, thereby solving this problem, but few plant and animal cell isolation problems have ready-made selective markers.

Viable-cell separation processes applied to this same problem can isolate 1,000 desired cells in a single step starting with 10_7 cells. For example, if an optical marker is available, clonogenic cells can be collected by sorting each desired cell into a single well with a cell sorter, resulting in 10 microwell plates with a selected, clonogenic cell in each well.   Similarly, if surface markers exist, selection with an affinity ligand is possible; if surface charge differences exist, electrophoretic isolation will yield a bulk suspension of selected cells in a few hundred μL of liquid; etc.

Hybrid (Electrofused) Cells.   In a rigorous statistical study of the rescue of desired hybridoma cells during monoclonal antibody production, Adamus et al. (18), found it necessary to perform two serial clonal selections by performing serial dilutions in 4 96-well microtiter plates at each step using 0.2 - 10 cells per well.   A physical process that can accomplish the same goals automatically would obviously be helpful as hybridoma technology is one of the most intense users of cell selection.

Preliminary research has indicated that pairs of cell types can be chosen with differing electrophoretic mobility and that heterodikaryons of such cells have an intermediate electrophoretic mobility ([19]). The use of a physical property averts the need for a selectable genetic marker in the heterokaryons. Thus it is no longer necessary to fuse only cells that have been genetically manipulated to possess markers that are lethal in hybrid selection medium.

Non-clonogenic Cells. Dispersed cells from plant and animal tissues contain very few cells capable of proliferation, but every tissue contains numerous types of cells. Physical processes that separate cell types from one another are required for the performance of functional studies or studies of composition.

Biochemical study. One of the major goals of cell separation technology is the isolation of adequate numbers of cells for biochemical study. The goals of biochemical analysis of isolated cell subpopulations are too numerous to list, but a few examples will illustrate the importance of this goal.

Cell subsets for transplantation. Cell subsets for transplantation can be isolated by physical methods, providing populations that can be transplanted in the absence of antigen-presenting cells, which enhance the host's immune rejection response, for example. Other types of undesired cells can be eliminated from transplant populations, such as prolactin-producing cells of the pituitary. In a classical example, graft-vs.-host cells have been removed from bone marrow cell populations, thereby enhancing host survival ([20],[21]).

Productive from Non-productive Bioreactor Cells. In modern bioreactor technology cells are often cultured to produce products specified by foreign genes. Cells with foreign genes suffer from an additional metabolic burden that usually slows their rate of multiplication relative to that of their counterpart cells without foreign genes. Therefore, cells that have lost foreign genes can take over a bioreactor culture in a few generations, so strategies must be used to rid bioreactors of non-productive cells.

High-cell-density bioreactors also tend to accumulate dead cells, which are also non-productive. A procedure for continuously or regularly ridding bioreactor cultures of dead cells is also useful, as dead cells can have various adverse affects on reactor productivity, such as secreting products that poison the live cells or their products, consuming nutrients from the medium uselessly, etc.

Preparation of Pure Materials. When an intracellular product is made by a small fraction of the total cells in a population, such as an animal tissue, it is not always useful to isolate the product from the total tissue, but instead, from a subpopulation of cells from that tissue.

Study of the processing of metabolites in organs. Different cell types in a given tissue possess different metabolic pathways, including pathways of hormone and vitamin transformation. The relationship between metabolic steps and functional cell types cannot always be established by histochemistry. While histochemistry can often establish the presence or absence of certain enzymes, in cannot provide enzyme kinetic data or enzyme characterization data.

Evaluation of separation methods

Resolution. The quantitation of resolution tends to follow the paradigm of multi-stage extraction or adsorption (chromatography).

$$R = (x_2 - x_1)/2(\sigma_1 + \sigma_2) \tag{1}$$

where R = "resolution" x_2 and x_1 are the migration distances of any two separands, and σ_1 and σ_2 are their respective standard deviations.

The inverse view is also useful. In sedimentation, in which resolution depends on the standard deviation of the distance sedimented and electrophoresis, in which it depends on the standard deviation of mobility, and in flow sorting, where it depends on the standard deviation of optical fluorescence, for example --all cases in which the moments of the appropriate measurement are known -- the inverse of the "Theoretical Stages" is a useful variable. This is the coefficient of variation ("CV") or "relative standard deviation", in any case the ratio of standard deviation to the mean. For example,

$$CV = \sigma_1/x_1. \tag{2}$$

There are usually physical constraints on resolution. The reciprocal of maximum number of fractions that can be collected, here termed number of separation units, NSU, cannot exceed the CV. This measure of "geometrical resolution" is defined strictly in terms of the geometry of the system. In a typical cell harvesting process, this would be determined from

$$NSU = \text{volume of separator/volume per fraction.} \tag{3}$$

In flow sorting, it works out that NSU = number of cells sorted and can be very large (more than 1 million). In some cell electrophoresis procedures, NSU can be as small as 12. Sometimes these fractions are further pooled into 4-6 suspensions, reducing NSU to 4-6. These numbers should be considered in comparison with chromatography, for example, where NSU = 10,000 is common (this number relates to the procedure for fraction collection and is not the same as the number of theoretical plates).

Purity. Purity is defined as the proportion of a separated fraction that consists of the desired product (cell type 1) among all cell types, i:

$$\text{Purity} = x_1/\sum_i x_i \, , \qquad\qquad (4)$$

where x_i's are cell concentrations. In a typical separation a trade is made between purity and recovery. The highest purity is found in one (possibly two) fraction(s). If these contain only 1/3 or less of the product, adjacent fractions can be pooled with the central fraction, thereby increasing recovery but reducing the average purity to

$$\text{Purity} = \{\sum_j (x_1/\sum_i x_i)V_j\}/\{\sum_j V_j\} \qquad\qquad (5)$$

for all cell types i in j pooled fractions.

One cannot assess "purification" unless one knows the purities of both the separated fractions and the initial suspension of cells. When reports state that certain separated fractions contain activity or "purified" cells without giving the level of activity or concentration of cells in the starting suspension, it is impossible to know whether or not a purification took place. The fact that different fractions from a separation contain different concentrations of separands is almost inevitable and is not necessarily evidence for any degree of purification. This point may not be appreciated when yields are not also calculated.
        There are many means of assessing purity. One can assess a functional parameter and express purity as a functional unit, i.e., secretion of a hormone, activity of an enzyme, ability to kill neoplastic cells etc., per "purified" cell. It is important to be aware that purity, as assessed morphologically or as expressed as proportions of cells with given phenotypic characteristics (labelling with a monoclonal antibody, containing phagocytosed bacteria, or whatever) is almost never directly related to purity expressed as average amount of function/cell. Let us consider an example that demonstrates the importance of microscopic examination. One might have a fraction number 12 that contains 100% macrophages with one or two ingested bacteria per cell. By this standard, the purification has been absolute, and the macrophages that could not ingest the tested bacterium have been partitioned into different fractions from the fraction with 100% macrophages that contain bacteria. It is possible that the same procedure for fractionation could yield a fraction 6 with (a) 25% macrophages that contain an average of 50 bacteria/ macrophage and (b) 75% macrophages that were unable to ingest bacteria. Let us further assume that the bacteria have not been killed and that the assay of purity is to lyse the macrophages, culture the bacteria, and count the number of bacterial colonies obtained. If we invent a situation in which all bacteria grow, one would expect to get 12.5 bacterial colonies/cell from fraction 6 and

1.2 bacterial colonies/cell from fraction 12. If no attempt were made to assess each cell individually, either morphologically or electronically with the appropriate labels, one might erroneously conclude that macrophages with the capacity to ingest bacteria were most highly purified in fraction 6. If the investigator were not to examine the purified cells in this model, it is also conceivable that the described model would show purification of presumed macrophages that contain bacteria in fractions that contain only masses of bacteria aggregated with large fragments of acellular debris.

It is difficult to overstate the value of photomicrographs of purified cells. Photomicrographs are valuable data in that they tell the reader (a) what kind of purification was accomplished and (b) how satisfactory the criteria were for the evaluation of the purification. The latter point may seem trivial; however, on occasion, unknown to the investigator, the quality of the preparations were quite inadequate to justify the conclusions that were drawn. Most descriptions of previously unreported methods for cell separations should include photomicrographs (22,23). These photomicrographs should be taken at a magnification that permits the inclusion of enough cells to convince the reader that a homogeneous fraction was obtained. A photomicrograph of one cell will always show 100% purity. It is not difficult to find fields with a few cells of the same type in most suspensions of cells prior to the separation of cells.

Bands that can be visualized or photographed in density gradients tells one nothing about purification and little about relative concentration. The ability of particles to scatter light and the resultant turbidity caused by a particle is a function of size. The presence of visible bands can often be taken as evidence for the aggregation of particles (organelles) or cells. Visible bands may occur at the location in a gradient where cells are present at the highest concentrations, since cells tend to aggregate more at higher concentrations; however, concentration is not the only determinant of the propensity of cells to aggregate, and the highest concentration of cells and/or the modal location of a particular population of cells in a gradient may be physically remote from the grossly visible bands caused by the aggregation of cells that have aggregated. If heterogeneous groups of cells were aggregated before they were layered over a gradient for isopycnic centrifugation, (a) one will reliably get grossly visible "bands" in the gradient and (b) the heterogeneous nature of the aggregates precludes significant purification of homogeneous subpopulations from cells that are aggregated.

In the purification of materials other than cells it has been conventional for years for investigators to keep a balance sheet that shows where all separated materials went. It is important to present some form of quantification of recovery and a graph that shows the locations of all cells at the end of the separation. It is impossible to assess critically the value of a cell separation in the absence of such data.

Viability. It is usually possible to contrive separation conditions that do not kill living cells. In some cases, ingenuity is required to minimize shear forces, eliminate toxic chemicals (including certain affinity ligands), incorporate physiologically acceptable buffer ions, maintain osmolarity, control the temperature, etc. Nearly every 'cell separation process is in danger of introducing conditions that kill cells, and additives such as glycine, albumin, serum, and neutral polymers are frequently used in cell separation systems.

The measurement and definition of "viability" may be complex, difficult, and controversial. The fact that cells are judged to be viable by one or more criteria does not guarantee that they (a) were not injured by the procedure for cell separation or (b) can perform all of the functions that are generally associated with viability. For example, some forms of zonal rotors are equipped with rotating seals. In the case of one type of these rotors, the shear forces experienced by even very "tough" cells in passing through a rotating seal were lethal. After passing through one of the rotating seals that is standard equipment with a particular zonal rotor, Ehrlich ascites tumor cells appeared morphologically intact, excluded trypan blue, and exhibited time-lapse motility; however, they ware not tumorigenic in syngeneic mice even when tested with a thousand-fold multiple of the usual tumorigenic dose (Pretlow et al., unpublished).

Capacity. The capacity of a cell separation method can be measured in number of cells processed per hour. Significant scale-up above 10^7 cells/hour is seldom practiced, partly owing to the availability of cells and partly owing to the cost of producing larger numbers of certain cell types that need to be separated. Cells with infinite reproductive capacity, such as yeast and bacteria, can usually be maintained and cultivated as pure populations, with the major exception being cultures with a high rate of mutagenesis caused by the loss of recombinant plasmid (24).

Relationship to Biological Function. Only in the cases of affinity methods and flow sorting are cell separation procedures based on the biological function for which the cells are being separated. In the cases of sedimentation, field-flow and electrokinetic methods, physical properties must be fortuitously linked to physiological/biological function.

Convenience. In the labor-intensive arena of biomedical research convenience may come in one of two forms: a simple process that requires very little engineering skill or physical manipulation or a complex process that has been highly automated by an expensive machine. These two extremes are perhaps represented by 1-g sedimentation on the one hand and by high-speed flow sorting on the other. Generally, biomedical researchers who do not wish to dedicate excessive amounts of manpower to separation process development seek convenience in a separation process, possibly to the exclusion of other attributes.

Cost.   The cost of a separation process includes capital equipment (apparatus), reagents and labor.   The economics of investing in separation technology are highly dependent upon goals.   A frequently repeated process, for example, is better done by an automated system and is capital-intensive.   A rarely performed, intellectually demanding procedure would be labor-intensive.   And a process that requires large quantities of expensive affinity reagents (such as antibodies) is expendables-intensive and would not be used to separate large quantities of separand without provisions for recycling ligands or reducing their cost.

Methods of cell separation

Optical and electronic sorting

General   characteristics.   Flow   cytometry   (FCM)   is   the performance of measurements on cells as they flow past a sensor, the signal of which can be translated into an electrical impulse upon which quantitative measurements (height, area, width, rise time, etc.)   can   be   made   and   tabulated.   Flow   sorting   consists   of transporting cells, single file, through a nozzle into a stream of solvent (usually physiological saline) so the cells can be monitored electronically ("Coulter volume") or optically one at a time.   The resulting electronic signal is used to charge or not charge a droplet containing a specific cell.   This drop, if charged, is deflected by a DC electric field into a collection vessel.   See Figure 1 in chapter 2.   It is possible to collect up to 5 subpopulations this way.   In addition to deflecting charged drops containing cells it has recently become possible to manipulate single cells in suspension by optical pressure or optical trapping.   This exciting new development is described in the chapter by T. Buican (25).

Flow sorting separates cells one at a time.   Thus resolution depends   on   the   intrinsic   optical   property   of   the   cell   that   is measured   as   selection   criterion.   This   property   is   always distributed, and CV = 0.20 can occur frequently.   In the case of narrowly distributed properties, such as DNA content, CV as low as 0.01 has been achieved.   Thus the resolution of the sorting process depends on the CV of a cell property in the starting population.

Applications.   Flow cytometry (FCM) has found extensive use in clinical immunology, and its maturity as a routine technology (vs. a basic research tool) has been proven in this field.   A limited number of applications in tumor pathology and clinical hematology have also become routine and accepted in the respective professions.   The flow cytometer has become a basic research tool in several areas of cell biology, pathobiology, and toxicology, to name a few examples, and a few thousand instruments are now in operation around the world.

Metabolic characterization of cells suspended from solid tissues can   be   performed   using   blue   autofluorescence   (an   indicator   of NADH/NADPH levels), rhodamine 123 staining (mitochondrial activity) and pyronin Y staining   (RNA content). These markers have been exploited (26) in the characterization of airway epithelial cells isolated by centrifugation and non-invasive flow sorting. The combination of these techniques made it possible to follow differentiation pathways in heterotopic tracheal grafts of pure populations of basal and secretory cells.

The characterization of messenger RNA being made by cells in vivo is possible through in situ hybridization using cloned nucleic acid probes complementary to specific mRNA. Highly radioactive probes, either synthesized or nick-translated using ^{32}P-labeled nucleotides, can be used in "Northern" blots and cell radioautography. Nucleic acid probes can also be tagged with fluorescent or immunoenzymatic labels or with biotin, and methods have been developed for applying such labeled probes to whole cells in suspension. These cells can then be evaluated by flow or image cytometry.

Non-cellular applications are also possible; beaded media, droplets undergoing phase separation, precipitates, standard test particles, and other non-living components of bioprocess technology can be assessed by FCM.

Flow Sorting of Cells and Chromosomes. Most early flow cytometry experiments were confined to the study of cells grown in vitro, immunological and hematological cells, and certain tumor cells. Improved cell dispersal methods have broadened the applicability of flow cytometry, even to include retrospective analysis of DNA distributions in nuclei prepared from tissue cells embedded in paraffin (27). Flow karyotyping, the study of isolated metaphase chromosomes in suspension by flow cytometry, has advanced to a high level of sophistication, and the effects of genotoxic agents are detected in the form of modified fluorescence intensity distributions. If there have been translocations that have been replicated, for example, these will appear as new peaks in the flow karyogram. If there has been extensive fragmentation, the "baseline" between chromosome peaks will rise. Mutant gene products can also be sought by flow cytometry. The preparation of chromosomes by combined sedimentation and flow sorting methods to produce enough material for genetic analysis is detailed by Albright and Martin in this volume (28).

By instrumenting the flow cytometer to detect "rare events" and by using a highly specific fluorescent antibody stain it is possible to detect 1 cell with a mutated surface protein per few million circulating erythrocytes. This approach was applied by Bigbee, Langlois and Jensen (29) to demonstrate that variant human erythrocytes expressing neither the M nor N glycophorin A gene occur more frequently in individuals bearing defective repair genes or exposed to genotoxic agents. The routine detection of 1 cell in 10,000 is presented by Leary et al. in this volume (30).

Bioprocessing applications. To date, this method has found important but limited application to bioprocess engineering and bio-processing research. Many bioprocessing research projects entail the study of living cells: bacteria, yeast, other fungi, plant cells, and animal cells. The study of each from the bioprocessing perspective has its own problems relative to sensitivity, wavelength, pulse analysis, and data analysis. The full range of cell types to which FCM can be applied in bioprocessing has not been explored. For example, the monitoring of viability, kinetics, metabolism, and production by single cells in a suspended-cell reactor can be effectively performed by FCM. It has become a chosen monitor of gene expression in eukaryotes in molecular biology and biotechnology. Bailey, Ollis, and others have used FCM in the verification of biomass models in cell bioreactors (31). However, the number of bioprocess researchers using FCM today is still minuscule. This is surprising in view of the ability of the flow cytometer to measure nearly all of the single-cell properties that are critical to cell bioreactor function.

Cell growth can be measured by the direct enumeration of cells by electronic or optical counting of cells in a specified volume of medium. Cell viability can be evaluated on the basis of fluorescein diacetate staining; fluorescein diacetate enters cells and is de-acylated to become fluorescent. Dead cells do not retain the dye while viable ones do (32). Similarly, dead cells admit propidium iodide through defective membranes, and this stains all nucleic acids in the cell (33). Thus two or more methods are available to determine viable cell count, and either live or dead cells can be stained. Fluorescent antibodies reacting with mouse Ig are commercially available, and these can be used to measure surface or internal Ig, depending on the method of cell preparation. Brief staining of live cells with fluorescent anti-Ig stains only surface Ig, while internal Ig is also stained in alcohol-formalin-treated cells. Cell size can be measured on the basis of the resistive impulse volume ("electronic", or "Coulter" volume), forward angle light scatter (0-2 degrees), and pulse-height independent optical (scatter or fluorescence) pulse width measurement (34). Position in the cell cycle can be determined on the basis of DNA content, which is measured using fluorescent staining with propidium iodide (following alcohol fixation and RNAase treatment) or with the DNA-specific dye Hoechst 33258 (UV illumination). The level of "reduced nucleotide" (NADH + NADPH) can be measured in each cell on the basis of "autofluorescence" (35) stimulated at 365 nm. This is a very important index of the metabolic (redox) state of the cell and can be correlated with cell kinetics, product release, and redox electrode measurements in the broth.

The ability of flow cytometers to make measurements on individual bacteria has improved over the last several years (36). With a modest mercury arc lamp, adequate numerical aperture, and appropriate staining (mithramycin plus ethidium bromide) it is possible to measure DNA in bacterial cells, thereby assessing the state of bioreactor cultures. Bacterial cells, being small, scatter large amounts of light at 90 degrees. Adequate sensitivity using antibody staining requires the use of special amplification techniques, such as enzyme-coupled avidin and biotin derivatives of primary antibodies. Most dye ligands used in fluorescence flow cytometry are compounds of traditional aromatic dyes, such as fluorescein, rhodamine, Hoechst compounds or carbocyanines, but efforts at achieving increased sensitivity and reduced spectral overlap by the use of rare-earth compounds are detailed by Vallarino and Leif in this volume (37).

Sedimentation. Like most physical methods, sedimentation separates cells on the basis of physical properties irrespective of their biological function. However, the density and radius of numerous eukaryotic cell types are consequences of their function.   For example, heavily granulated cells, such as granulocytes and hormone-secreting cells of the anterior pituitary are large and dense, owing to their cytoplasms being filled with high-density granules that share many properties with protein crystals.

The general equation of motion that is exploited in sedimentation is

$$v = dx/dt = 2(\rho - \rho_o)a^2\alpha/9\eta \tag{6}$$

where v is the terminal velocity achieved when the Stokes drag force on a sphere, $6\pi\eta av$, exactly balances the acceleration, α, and buoyancy, $(\rho - \rho_o)$, forces. The particle (cell) radius is a, its density is ρ, and η and ρ_o are viscosity and density of the medium, respectively.

In 1-g sedimentation the acceleration is g = 9.8 m/s^2, and this has been used in bone marrow and peripheral blood cell separations (38). In typical 1-g applications, a density gradient $\rho_o(x)$ is used, so dx/dt is a function of x and therefore of time. In simple centrifugation the acceleration vector is ω^2x, where x = distance of the separand particle from the center of rotation. Clearly dx/dt is a function of x in centrifugation with or without a density gradient. Centrifugation in the absence of a density gradient distributes cells radially according to their radius and differential density. This process has been defined as "differential centrifugation", meaning "the separation of homogeneous mixtures of heterogeneous particles by centrifugation in the absence of a density gradient with just sufficient force to permit a crude purification of the most rapidly sedimenting particles on the bottom of the centrifuge tube and of the most slowly sedimenting particles that remain in suspension (39).

Three variants can be used to extend the separation process: one is to create a density gradient that results in $(\rho - \rho_o)x/\eta$ = const., which gives v = constant. Such a density gradient is an "isokinetic gradient", as discussed in the chapter by Pretlow and Pretlow (40). Also, fluid can be pumped inward at velocity dx/dt for a particular separand, so that slowly separating separands are pumped back out the top, and rapidly sedimenting components continue out the "bottom" (outer radius) of the centrifuge -- this is elutriation and is detailed in the chapter by Keng (41). Isopycnic sedimentation is an equilibrium process. Particles sediment, either at unit gravity or in a centrifugal field, in a density gradient until $\rho = \rho_o$, at which time sedimentation (buoyancy) stops. Fractions are then collected on the basis of particle density.

Affinity Adsorption. From the standpoint of separation thermodynamics, affinity adsorption is an <u>equilibrium</u> process. It has traditionally been practiced in three ways: single-stage batch desorption, multi-stage adsorption and desorption, and continuous desorption, as in chromatography. Most applications to cells are single-stage batch separations. The selection and evaluation of adsorption media for cells is discussed in the chapter by Kataoka (<u>42</u>).

Few processes separate cells on the basis of their intrinsic magnetic properties. However, highly selective, high-capacity hydrodynamic capture methods have been developed using magnetic fields, such as magnetic filtration, high-gradient field separation, and magnetic flotation (<u>43</u>). Affinity adsorption of magnetic particles by cells is the source of specificity in magnetic cell separations. Magnetically enhanced affinity methods are becoming popular in cases in which the attachment of cells to a macroscopic surface may be undesirable due to interference by non-specific adhesion or untimely activation of cellular processes such as blastogenesis or cytokine production. An example of progress in cell isolation by magnetically enhanced affinity methods is presented in the chapter by Powers and Heath (<u>44</u>), and its potential commercialization is discussed by Liberti (<u>45</u>).

One of the major problems in magnetic capture is the occurrence of undesired adventitious interactions of magnetic particles owing to an attractive-only force that cannot be switched off. The result is cell aggregation and the aggregation of microbeads in the absence of cells or a magnetic field. The general approach to solving this problem has been the development of paramagnetic or superparamagnetic microspheres with affinity ligands. Superparamagnetic particles possess no magnetic moment unless they are in an inhomogeneous magnetic field of high gradient. Such particles have been made of ferritin and dextran or other beaded media (<u>46</u>, <u>47</u>).

Magnetic capture and affinity binding are both <u>equilibrium</u> processes. Equations of motion therefore apply only to the rate at which equilibrium is achieved. Free energies of these equilibria depend on the magnetic dipole moment of each particle, the number of particles adsorbed per cell, strength of the magnetic field gradient, and the ligand affinity constant.

Biphasic Extraction. Biphasic extraction is one of the most popular purification methods used in the chemical industry today. It has found limited popularity in bioprocessing owing to the damaging effects of organic solvents on biomolecules and cells, whereas aqueous two-phase systems, due to their high water content, are biocompatible (<u>48</u>,<u>49</u>). Moreover, these systems are reported to have provided stability to biologically active substances, such as enzymes (<u>50</u>). Due to their similar physical properties, immiscible aqueous phases do not separate rapidly in large volumes, as in production-scale purifications. Despite some 800 papers on this subject (<u>4</u>), large-scale commercial applications are not widespread. The high cost of lower-phase polymers is another deterrent to its widespread

use. Cost containment has been recently affected by the introduction
of low-cost polymer aqueous phase systems (51).

In practice, multistage extractions are performed to achieve
high-resolution separations (52). The physical problems associated
with biphasic extraction research can be divided into two major
categories, although they bear certain thermodynamic similarities:
phase separation --the formation of two phases from a dispersion, and
partitioning -- the preferential transfer of a separand into one
phase.

Phase separation. When two polymers A and B are dissolved in
aqueous solution at concentrations that cause phase separation an
upper phase forms that is rich in A and poor in B, and a lower phase
forms that is rich in B and poor in A. Typically A is polyethylene
glycol (PEG), considered a relatively hydrophobic solute, and B is
dextran or a similar polysaccharide. B can also be a salt at high
concentration. The phase separation process is described by a two-
dimensional phase diagram, such as in Figure 2, in which high
concentrations of A and B cause the formation of top-phase solutions
with compositions given by points in the upper left and bottom-phase
solutions with compositions given by points in the lower right. Each
combination of A and B falls on a "tie line" connecting the resulting
top and bottom phase compositions at equilibrium. Higher
concentrations of polymers result in longer tie lines on the phase
diagram. Any composition that lies on a tie line will result in the
same equilibrium compositions, and the ratio of the volumes of the
two phases is related to the position of the initial composition on
the tie line. The curve that forms the envelope connecting the ends
of the tie lines, the "binodial", shown in Figure 2, separates the 1-
phase and 2-phase regions on the diagram. As an example, one two-
phase extraction system is described by the locations of the
encircled points on the phase diagram of the PEG/dextran/water system
at 25°C shown in Figure 2. The top and bottom phase densities of this
system are 1.0164 and 1.1059 g/cm^3, respectively, and the
corresponding viscosities are 0.0569 and 4.60 Poise.

Such phase diagrams are strictly experimental; however, they
represent thermodynamic equilibria, and they should be predictable on
the basis of thermodynamic principles. Cabezas et al. (53), chose
to apply statistical mechanics via the solution theory of Terrell
Hill (54). At equilibrium the chemical potential of each substituent
(typically dextran, PEG and water) is the same in the top and bottom
phase, and the chemical potential of each can be determined from
their fractional molalities m_i and their osmotic virial coefficients
C_{ij} by

$$\Delta\mu_2 = -RT[\ln m_2 + 2C_{22}m_2 + 2C_{23}m_3] \tag{7a}$$

$$\Delta\mu_3 = -RT[\ln m_3 + 2C_{23}m_2 + 2C_{33}m_3] \tag{7b}$$

$$\Delta\mu_1 = -RT[m_2 + m_3 + C_{22}m_2{}^2 + 2C_{23}m_2m_3 + C_{33}m_3{}^2] \tag{7c}$$

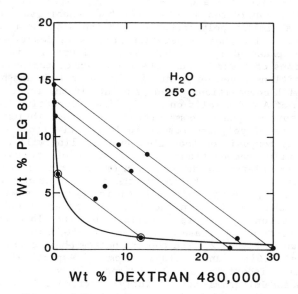

Figure 2. Phase diagram of the dextran–water–polyethylene glycol system showing tie lines and binodial curve. The vertical limb of the binodial curve gives compositions of upper phases and the horizontal limb gives compositions of lower phases at equilibrium. (Reproduced with permission from reference 61. Copyright 1990 by Marcel Dekker.)

where subscripts 1, 2 and 3 correspond to water, PEG and dextran, respectively. The same relationships apply to any pair of polymers, but not to polymer-salt combinations that form two phases, as electrostatics must be added to account for the chemical potentials of salts. Thermodynamic analyses that account for the interactions of ions have been recently achieved (55).

The osmotic virial coefficients C_{ij} are for polymers and constitute additional unknowns in equations (7). These can be derived from group renormalization theory as applied to polymer solutions (56) by using monomer-monomer interaction coefficients b_{ij}:

$$C_{22} = b_{22}N_2^{3\nu 2}[1 + (2/9)\ln(M_{w2}/M_{n2})] \tag{8a}$$

$$C_{33} = b_{33}N_3^{3\nu 3}[1 + (2/9)\ln(M_{w3}/M_{n3})] \tag{8b}$$

$$C_{23} = b_{23}\{N_2^{3\nu 2}[1 + (2/9)\ln(M_{w2}/M_{n2})][N_3^{3\nu 3}[1 + \tag{8c}$$

$$(2/9)\ln(M_{w3}/M_{n3})]\}^{1/2}$$

where C_{23} is based on an empirical application of the geometric mean rule. At equilibrium the chemical potential of each constituent in the top phase is equal to its chemical potential in the bottom phase. These equalities result in a system of equations with the same number of equations as unknowns which can be solved for m_2 and m_3 to obtain equilibrium concentrations in both phases. These concentrations have been found to successfully predict phase diagrams, including the one shown in Figure 2 (53).

In addition to these equilibrium phenomena, rates of phase separation ("demixing") are of technical interest (57). When two polymers are dissolved in aqueous solution at concentrations that cause phase separation, certain dissolved ions such as phosphate are unequally partitioned between the phases (58) leading to a Donnan potential across the interface (59) and an electrokinetic (zeta) potential at the interface (60). As a consequence of the latter, phase demixing can be hastened by the application of an electric field (61).

Partitioning. Partitioning of molecules and cells between phases during demixing is also considered a thermodynamic process and not a rate process. This means that scale up under relatively non-hostile conditions should be feasible, and this is the main promise of biphasic aqueous extraction as a purification method. Partitioning was originally modelled by Brønsted (62), who noted that, at the very least, partition coefficient, K, should depend on the molecular weight of the separand

$$K = \exp(\lambda M/k_B T) \tag{9}$$

but there are at least 4 properties that determine a molecule's partition coefficient: molecular weight, hydrophobicity, charge density, and binding affinity. Brooks and others (49, 63, 64, 65) consider the surface area, A_M (which does depend on molecular weight)

of a partitioning molecule and its interfacial free energy per unit
area $\Delta\gamma$, as the significant measurable variables in determining
partition coefficient:

$$K = \exp(\Delta\gamma A_M/k_B T). \tag{10}$$

While such a relationship seems valid for solute molecules, it does
not satisfactorily describe the partitioning of particles, such as
cells. In one view, equation (10) would only be satisfactory if T
could be set to 10^4 - 10^5 °K. Such a high $k_B T$ implies randomizing
effects due to non-thermal forces, such as gravity (63-65).
Furthermore it is necessary to account for electrokinetic transport
as a means of reaching equilibrium when an electrical potential, $\Delta\Psi$,
between the two phases exists (48). A particle can interact with
both phases simultaneously, so the difference between its interfacial
free energies with respect to the top phase, γ_{pt} and with respect to
the bottom phase, γ_{pb} , which is $\Delta\gamma$, is a determinant of partition
coefficient. The interfacial free energy between top and bottom
phase γ_{tb} plays a role in excluding particles from the interface. The
suggested overall thermodynamic relationship for partitioning out of
the interface is

$$K = B\exp\{-\gamma_{tb}A[1 - (\Delta\gamma + \sigma_e\Delta\Psi/\gamma_{tb}]^2/4k_B T\}, \tag{11}$$

where B is a constant of proportionality. When salts, especially
salts of sulfate and phosphate, are dissolved in PEG-dextran
solutions, an electrochemical potential is developed between the
phases (58,59). Arguments derived from equilibrium thermodynamics
suggest that this could be a Donnan potential developed by the
unequal partitioning of ions between the two phases (59), and this
potential can drive cells into the upper (more positive) phase on the
basis of cell surface charge density ("zeta potential").

Field Flow Fractionation. This technique, investigated and developed
mainly by J. C. Giddings and co-workers (see chapters 9 and 10 by
Giddings (66) and by Bigelow et al. (67)) and abbreviated FFF, is
defined as any separation method in which a transverse field is
imposed on dissolved or suspended separands as they flow through a
chamber. The transverse field may be a pressure drop imposed through
porous chamber walls (flow field), an electric field, a centrifugal
acceleration field (steric FFF) an adhesion force at the chamber wall
(67), or simply the shear rate due to the velocity gradient imposed
by Poiseuille laminar flow. Typically, in applications to cells, a
suspension of mixed cells flows into a chamber that is 100 - 300 μm
thick, a few cm wide and several cm long. The thickness of the
chamber establishes the strength of transverse field that can be
applied (usually flow field), the width of the chamber establishes
its capacity, and the length of the chamber establishes, up to some
maximum length, the resolution of the separation.

Applications of field-flow separations to cells have not been
numerous. It is primarily a chemical separation technique. Recent
studies have shown quite clearly, nevertheless, that particles of

different sizes, differing by perhaps 1 μm in diameter, can be separated from one another with useful resolution. In another recent development, the separation of cells on the basis of their different adhesion strengths to the chamber walls has been accomplished (67). In a sense, this is a classical chromatographic technique, in which adsorption is the thermodynamic variable exploited to effect a separation. This method also has analytical as well as separative value; by knowing the shear rate of the flowing buffer required to remove adherent cells, the magnitude of the adhesion force can be evaluated.

Field-flow technology has not enjoyed widespread use in either laboratory or industrial cell separations. This is the case with several biophysical separation methods in which practitioners unfamiliar with the physical principles underlying a separation process are reluctant to exploit its efficiency owing to a phobia for physical/mechanical things. There is a widespread feeling that centrifuges and chromatographs are for biologists and biochemists while free electrophoresis, elutriation, field-flow fractionation and, to some extent, optical sorting methods are for physicists and engineers.

Electrokinetic Methods. Electrophoresis is the motion of particles (molecules, small particles and whole biological cells) in an electric field and is one of several electrokinetic transport processes. The velocity of a particle per unit applied electric field is its electrophoretic mobility, μ; this is a characteristic of individual particles and can be used as a basis of separation and purification. This separation method is a <u>rate</u> (or transport) process.

The four principal electrokinetic processes of interest are electrophoresis (motion of a particle in an electric field), streaming potential (the creation of a potential by fluid flow), sedimentation potential (the creation of a potential by particle motion), and electroendosmosis (the induction of flow at a charged surface by an electric field, also called electroosmosis). These phenomena always occur, and their relative magnitudes determine the practicality of an electrophoretic separation or an electrophoretic measurement. It is generally desirable, for example, to minimize motion due to electroendosmosis in practical applications. A brief discussion of the general electrokinetic relationships follows.

The surface charge of suspended particles prevents their coagulation and leads to stability of lyophobic colloids. This stability determines the successes of paints and coatings, pulp and paper, sewage and fermentation, and numerous other materials and processes. The surface charge also leads to motion when such particles are suspended in an electric field. The particle surface has an electrokinetic ("zeta") potential, ζ, proportional to σ_e, its surface charge density - a few mV at the hydrodynamic surface of stable, non-conducting particles, including biological cells, in aqueous suspension. If the solution has electrical permittivity ϵ (= 7×10^{-9} F/m in water), the electrophoretic velocity is

$$v = \frac{2\zeta\epsilon}{3\eta}\ E \qquad\qquad\qquad (12)$$

for small particles, such as molecules, whose radius of curvature is similar to that of a dissolved ion (Debye-Hückel particles), and

$$v = \frac{\zeta\epsilon}{\eta}\ E \qquad\qquad\qquad (13)$$

for large ("von Smoluchowski") particles, such as cells and organelles. η is the viscosity of the bulk medium. At typical ionic strengths (0.01 - 0.2 g-ions/L) particles in the nanometer size range usually have mobilities (v/E) intermediate between those specified by equations (11) and (12) (68).

Analytical electrophoresis of proteins and other solutes is performed in a gel matrix, because convection is suppressed; however, high sample loads cannot be used owing to the limited volume of gel that can be cooled sufficiently to provide a uniform electric field, and particulate separands as large as cells do not migrate through gels. Capillary zone electrophoresis, a powerful, high-resolution analytical tool (69), depends on processes at the micrometer scale and is not applicable to preparative cell electrophoresis. Therefore preparative electrophoresis must be performed in free fluid. The two most frequently used free-fluid methods are zone electrophoresis in a density gradient and free-flow (or continuous flow) electrophoresis (FFE or CFE). Other preparative methods, more suitable for molecular separations, are described in reviews by Ivory (70) and by Mosher et al. (71). The combination of CFE with complimentary methods is shown to be a powerful approach to the analysis of cells of the immune system in chapters 13 and 14 by Crawford et al. (72) and by Bauer (73).

Analytical and preparative cell electrophoretic methods were compared critically in a review by Pretlow and Pretlow in 1979 (74). The number of preparative methods alone has since grown to at least 13, and these are introduced, reviewed and compared quantitatively in Chapter 15 (75).

The electrophoresis of living cells imposes physical constraints on solutions that can be used for electrophoresis buffers. While maintaining low conductivity it is also necessary to maintain isotonic conditions for the cells. This is usually achieved by the addition of neutral solutes that are not harmful to cells, such as sugars. With a few notable exceptions, living cells do not tolerate temperatures above 40°C, so, thermoregulation designed to prevent natural convection also must account for the temperature sensitivity of the separands. These and other constraints are addressed in Chapters 12-15.

Summary

The chapters that follow indicate that the science and technology of
cell separation is becoming more diverse, that four principal methods
of cell separation are emerging and being applied to increasingly
diverse separation problems, that cell separation science and
technology continues to be a research area of its own as applications
increase, that each technology is increasing in scientific
sophistication, and that continuing dialogue between users and
developers of cell separation methods contributes to progress in this
important field.

Literature Cited

1. Cell Separation Methods Bloemendal, H. Ed.
    Elsevier/North-Holland Biomedical Press: Amsterdam, 1977.
2. Cell Separation. Methods and Selected Applications. Pretlow,
    T.G.,II; Pretlow, T.P., Eds.; Academic Press: New York, NY,
    1978-1987, Vol. 1-Vol. 5.
3. Methods of Cell Separation; Catsimpoolas, N., Ed.; Plenum
    Press: New York, NY, 1977, 1979; Vol. 1, Vol 2.
4. Partitioning in Aqueous Two-Phase Systems. Walter, H.; Brooks,
    D.E.; Fisher, D., Eds.; Academic Press: New York, NY, 1985.
5. Todd,P.; Plank, L.D.; Kunze, M.E.; Lewis, M.L.; Morrison, D.R.;
    Barlow, G.H.; Lanham, J.W.; Cleveland, C. J. Chromatography
    1986, 364, 11-24.
6. Hannig, K. Electrophoresis 1982, 3, 235-243.
7. Takayasu, M.; Maxwell, E.; Kelland, D.R. IEEE Transactions on
    Magnetics 1984, 10, 1186-1188.
8. Walter, H.; Coyle, R.P. Biochim. Biophys. Acta 1968, 165, 540-
    543.
9. Walter, H.; Krob, E.J.; Brooks, D.E. Biochemistry 1976, 15,
    2959-2964.
10. Schlossman, S.; Hudson, L. J. Immunol. 1973, 110, 313-315.
11. Lillie, R.S. Amer. J. Physiol. 1902, 8, 273-283.
12. Beijerinck, M.W. Zeitz. für Chemie u. Industrie der Koll.
    1910, 7, 16-20.
13. Plank, L.D.; Hymer, W.C.; Kunze, M.E.; Todd, P. J. Biochem.
    Biophys. Meth. 1983, 8, 273-289.
14. Heidrich, H.-G.; and Dew, M.E. J. Cell Biol. 1983, 74,
    780-788.
15. Kreisberg, J.I.; Sachs, G.; Pretlow, T.G.,II; and McGuire, R.A.
    J. Cell Physiol. 1977, 93, 169-172.
16. Hymer, W.C.; Hatfield, J. In Methods in Enzymology; Colowick,
    S.J.; Kaplan, N.O., Eds.; Academic Press: New York, NY, 1983,
    Vol. 103; pp. 257-287.
17. Hymer, W.C.; Barlow, G.H.; Cleveland, C.; Farrington, M.;
    Grindeland, R.; Hatfield, J.M.; Lanham, J.W.; Lewis, M.L.;
    Morrison, D.R.; P. H. Rhodes, P.H.; Richman, D.; Rose, J.;
    Snyder, R.S.; Todd, P.; Wilfinger, W. Cell Biophysics 1987,
    10, 61-85.

18. Adamus, G.; Zam, Z.S.; Emerson, S.S.; Hargrave, P.A.. In Vitro Cell Dev. Biol. 1988, 25, 1141-1146.
19. Saunders, J.A.; Beretta, D.; Todd, P.; Matthews, B.S.; Singer, D.W. International School of Biomembrane and Receptor Mechanisms, Cannizzaro, Italy, Sep-Oct, 1985.
20. Nordling, S.; Anderson, L.C.; Häyry, P. Science 1972, 178, 1001-1002.
21. Zeiller, K.; Hannig, K.; Pascher, G. Hoppe-Seyler's Z. Physiol. Chem. 1971, 352, 1168-1170.
22. Pretlow, T.G., II; Weir, E.E.; Zettergren, J.G. Int. Rev. Exp. Pathol. 1975, 14, 91-204.
23. Pretlow, T.G.; Jones, C.M.; Pretlow, T.P. Biophys. Chem. 1976, 5, 99-106.
24. Davis, R.H.; Lee, C.Y.; Batt, B.C.; Kompala, D.S. This volume.
25. Buican, T. This volume.
26. Johnson, N.F.; Hubbs, A.F.; Thomassen, D.G. In Multilevel Health Effects Research: from Molecules to Man. Park, J.F.; Pelroy, R.A., Eds.; Battelle Press: Columbus, OH, 1989, pp. 135-138.
27. Kute, T.D.; Gregory, B.; Galleshaw, J.; Hopkins, M.; Buss, D.; Case, D. Cytometry 1988, 9, 494-498.
28. Albright, K.; Martin, J. This volume.
29. Bigbee, W.L.; Langlois, R.G.; Jensen, R.H. In Multilevel Health Effects Research: from Molecules to Man. Park, J.F.; Pelroy, R.A., Eds.; Battelle Press: Columbus, OH, 1989, pp. 139-148.
30. Leary, J.F. This volume.
31. Srienc, F.; Campbell, J.L.; Bailey, J.E. Biotech. Ltrs. 1983, 5, 43.
32. Rotman, B.; Papermaster, B.W. Proc. Natl. Acad. Sci. U.S.A. 1966, 55, 134-141.
33. Wallen, C.A.; Higashikubo, R.; Roti Roti, J.L. Cell Tissue Kinet. 1983, 16, 357-365.
34. Leary, J.F.; Todd, P.; Wood, J.C.S.; Jett, J.H. J. Histochem. Cytochem. 1979, 27, 315-320.
35. Thorell, B. Cytometry 1983, 4, 61.
36. Steen, H.B.; Skarstad, K.; Boye, E. Ann. N. Y. Acad. Sci. 1988, 468, 329-338.
37. Leif, R.C.; Vallarino, L.M. This volume.
38. Miller, R.G.; Phillips, R.A. J. Cell Physiol. 1969 73, 191-201.
39. Anderson, N.G. National Cancer Institute Monograph 1966, 21, 9-40.
40. Pretlow, T.G.,II; Pretlow, T.P. This volume.
41. Keng, P.C. This volume.
42. Kataoka, K. This volume.
43. Owen, C.S. Cell Biophys. 1986, 8, 287-296.
44. Powers, F.; Heath, C.A.; Ball E. D.; Vredenburg, J.; Converse, A. O. This volume.
45. Liberti, P.A.; Feeley, B.P. This volume.
46. Owen, C.S. IEEE Trans. Magn. 1982, 18, 1514-1516.
47. Kronick, P.L.; Campbell, G.L.M.; Joseph, K. Science 1978, 200, 1074-1076.

48.   P.-Å. Albertsson. Partition of Cell Particles and Macromolecules, 2nd Ed. Wiley-Interscience, New York, NY, 1971.

49.   P.-Å. Albertsson. Partition of Cell Particles and Macromolecules. Third Ed., John Wiley and Sons, New York, NY, 1986.

50.   Shanbhag, V.P. Biochim. Biophys. Acta 1973, 320, 517-527.

51.   Szlag, D.C.; Giuliano, K.A. Biotechnol. Techniques 1988, 2, 277-282.

52.   Van Alstine, J.M.; Snyder, R.S.; Karr, L.J.; Harris, J.M. J. Liquid. Chromatog. 1985, 8, 2293-2313.

53.   Cabezas, H. Jr.; Evans, J.; Szlag, D.C. In Downstream Processing and Bioseparation. Recovery and Purification of Biological Products. Hamel, J-F. P.; Hunter J.B.; Sikdar, S.K., Eds.; ACS Symposium Series 419, American Chemical Society: Washington, DC, 1990, pp. 38-52.

54.   T. L. Hill, T.L. J. Chem. Phys. 1959, 30, 93-97.

55.   Cabezas, H., Jr.; Kabiri-Badr, M.; Snyder, S.M.; Szlag, D.C. In Frontiers in Bioprocessing II; Sikdar, S.K.; Bier, M.; Todd, P., Eds.; American Chemical Society, Washington, DC, 1991.

56.   Schafer, L.; Kappeler, C. J. de Phys. 1985, 46, 1853.

57.   Baird, J.K. In Proc. 5th Europ. Symp. on Materials Science under Microgravity. European Space Agency: Paris, Series ESA SP-222, 1984, pp. 319-324.

58.   Johansson, G. Biochim. Biophys. Acta 1970, 221, 387-390.

59.   Bamberger, S.; Seaman, G.V.F.; Brown, J.A.; Brooks, D.E. J. Colloid. Interface Sci. 1984, 99, 187-193.

60.   Brooks, D.E.; Sharp, K.A.; Bamberger, S.; Tamblyn, C.H.; Seaman, G.V.F.; Walter, H. J. Colloid. Interface Sci. 1984, 102, 1-13.

61.   Raghava Rao, K.S.M.S.; Stewart, R.M.; Todd, P. Sep. Sci. Technol. 1990, 25, 985-996.

62.   Brønsted, J.N. Z. Phys. Chem. A, Bodenstein Festband, 1931, 257-266.

63.   Brooks, D.E.; Bamberger, S.; Harris, J.M.; Van Alstine, J. Proc. 5th European Symp. Material Sciences under Microgravity 1984, ESA SP-222, 315-318. 64. Brooks, D.E.; Sharp, K.A.; Fisher, D. In Partitioning in Aqueous Two-Phase Systems, Walter, H.; Brooks, D.E.; Fisher D., Eds.; Academic Press, New York, NY, 1985, pp. 11-84.

65.   Van Alstine, J.M.; Karr, L.J.; Harris, J.M.; Snyder, R.S.; Bamberger, S.B.; Matsos, H.C.; Curreri, P.A.; Boyce, J.; Brooks, D.E. In Immunobiology of Proteins and Peptides IV. Atassi, M.Z., Ed.; Plenum Publishing Corp., New York, NY, 1987, pp. 305-326.

66.   Giddings, J.C.; Barman, B.N.; Liu, M.-K. This volume.

67.   Bigelow, J.C.; Nabeshima, Y.; Kataoka, K.; Giddings, J.C. This volume.

68.   R. W. O'Brien, R.W.; and L. R. White, L.R. J. Chem. Soc. Faraday II 1978, 74, 1607-1626.

69.   Jorgenson, J.W.; Lukacs, K.D. Science 1983 222, 266-272.

70.   Ivory, C.F. Sep. Sci. and Technol. 1988, 23, 875-912.

71.   Mosher, R.; Thormann, W.; Egen, N.B.; Couasnon, P.; Sammons,
      D.W.   In New Directions in Electrophoretic Methods, J.W.
      Jorgenson, J.W.; Phillips M. Eds.; American Chemical Society,
      Washington, DC, 1987, pp. 247-262.
72.   Crawford, N.; Eggleton, P.; Fisher, D.   This volume.
73.   Bauer, J.   This volume.
74.   Pretlow, T.G.,II; Pretlow, T.P.   Int. Rev. Cytol., 1979, 61,
      85-128.
75.   Todd, P.   This volume.

RECEIVED March 15, 1991

FLOW SORTING AND OPTICAL METHODS

Chapter 2

High-Resolution Separation of Rare Cell Types

James F. Leary[1], Steven P. Ellis[2], Scott R. McLaughlin[1], Mark A. Corio[1],
Steven Hespelt[1], Janet G. Gram[1], and Stefan Burde[1]

[1]Department of Pathology and Laboratory Medicine and [2]Division of
Biostatistics, University of Rochester, Rochester, NY 14642

Isolation of cells by fluorescence-activated cell sorting, while useful, has been of only limited value. It is not a good isolation method
for obtaining large numbers of cells. However, two recent developments, one in the technology of cell sorting and the other in the field
of molecular biology make cell sorting extremely powerful for
some applications. New high-speed (100,000 cells/sec) analysis
and sorting of rare (0.01 percent) cell subpopulations allows more
than 10,000-fold enrichments on the basis of multiple parameters,
something not readily attainable by other cell separation technologies. Also, original frequency information important for
studies in toxicology and genetics is not lost by this method.
Second, application of new polymerase chain reaction (PCR) technologies from molecular biology means that isolation of a single
cell may be sufficient to provide necessary material for enzymatic
expansion of cellular DNA or RNA to levels equivalent to 10^6 - 10^9
cells. Thus a single sorted cell may provide preparative amounts of
DNA or RNA for further characterization by standard molecular
biology methodologies.

Analysis and isolation of rare cell subpopulations are of interest to researchers
and clinicians in many areas of biology and medicine including: (a) detection of
somatic cell mutations (1) in mutagenized cells, (b) detection of human fetal
cells in maternal blood for prenatal diagnosis of birth defects (2,3), (c) detection
of CALLA+ cells (4), and (d) detection of minimal residual diseases (5). Conventional flow cytometer/cell sorters operating at rates below 10,000 cells/sec
require many hours to analyze and/or isolate cell subpopulations of low frequencies (e.g. 10^{-5} - 10^{-7}).

0097–6156/91/0464–0026$06.00/0

While analysis of rare cell subpopulations dates back to the 1970's, as for example in Herzenberg's early attempts to isolate fetal cells from maternal blood (*3*), rare-event analysis lacked the technology to make it a practical method for analysis and separation of rare cell subpopulations. There was an obvious need for faster cell processing speeds for analysis and sorting of rare cell subpopulations. Systems have been built to separate cells at rates of 15,000 - 25,000 cells/sec (*6*). While these systems employ newer methods and technological advances such as faster analog-to-digital converters and multi-stage buffering of incoming signals to achieve faster cell processing rates they retain the paradigm of the original flow cytometers/cell sorters, namely digitization and storage of information as correlated or uncorrelated data consisting of all signals from all cells. This paradigm, while important and necessary for some applications, particularly for processing of non-rare cell subpopulations, imposes severe and perhaps unnecessary restrictions on the analysis and sorting of rare cell subpopulations. A characteristic of "rare event analysis" is that most of the cells are not-of-interest. A simple but very important alternative paradigm is to classify signals as "of interest" or "not of interest" prior to digitization. It is then possible, using relatively simple circuitry, to count all cells for original frequency information but to only digitize information from cells "of interest" or from cells which cannot be reliably classified by this procedure. The benefits of such a paradigm are two-fold. First, the circuitry required to operate flow cytometers at rates of more than 100,000 cells/sec becomes simpler and less expensive and can be implemented on existing commercially available flow cytometers. Second, it reduces the problems of storing and analyzing data sets containing 10^7 - 10^9 cells by storing only data of interest or data about which the experimenter cannot be certain as to whether it must be stored for further analysis. Data classified by the system as "not of further interest" can be counted but not digitized and/or stored.

This chapter describes a method and apparatus for the multiparameter high-speed measurement of a rare subpopulation of cells amidst a larger population of cells with differing characteristics. A multiparameter hardware/software system (U.S. patent pending) was developed which, when attached to a multiparameter flow cytometer/cell sorter and microcomputer, allowed multiparameter analysis of cells at rates in excess of 100,000 cells/sec. This system is an outboard module which can, with minor modifications, be attached to any commercially available or home-built flow cytometer. It allows analysis of 100,000,000 cells in less than 15 minutes as opposed to more than 6 hours on the same instrument without this module. The system provides for high speed counting, logic-gating, and count-rate error-checking. Indirectly, by acting as a high-speed front-end filter of signals, the system can be used to control high-speed cell sorting.

Actual instrument dead-time depends on the pulse widths of the signals as well as delay lines, if used. The actual through-put rate is limited not by the signal and software processing times, but rather by the in-excitation-beam cell coincidence caused by asynchronous cell arrival times in the cell sorter or similar type of device.

Use of thresholds and logical gating from total and rare cell signals with other non-rare signals allows multiparameter rare-event listmode data (a record of all pulses for each rare cell and the total number of cells to obtain original frequency information vital to many applications) to be acquired reasonably both in terms of signal processing speeds and total amount of data to be stored by a conventional second-stage data acquisition system. Analysis of these multiparameter rare-event data also permit reduction or elimination of many "false positives", an important problem in the analysis and isolation of rare cell subpopulations (see **Figure 1**). This is achieved by further data processing techniques such as principal component/biplot analysis, provided by a second-stage out-board module.

Some basic problems of rare cell sorting

Specific labeling, no matter how it is defined, requires that the signal be greater than the "noise". The goal of specific labeling is then to improve the signal-to-noise (S/N) ratio to the point where it permits unequivocal identification of rare cells for cell sorting. In a study of Rh incompatibility (2) we previously demonstrated that indirect immunofluorescence labeling of Rh-positive cells could not be accomplished at the level of 0.1 percent rare cells due to an insufficient S/N ratio. However, using the same primary antibody but using a secondary antibody labeled with a fluorescent bead we were able to analyze and sort rare cells as low as .002 percent. Not only was the signal from the immunobead more than 200 times brighter than by normal indirect immunofluorescence, but the background "noise" non-specific binding was actually less, resulting in an outstanding S/N ratio for this application. This labeling approach was not as good for some other applications such as those involving cells which can phagocytize immunobeads. However, a multiparameter labeling approach can give even better results. Many false-positive cells can be eliminated by requiring the presence or absence of two or more fluorescent signals using either specific or non-specific antibodies labeled with different fluorescent colors. Addition of other intrinsic flow cytometric parameters such as cell size (by pulse width time-of-flight) or 90-degree light scatter, can significantly reduce the number of false-positive cells, thereby improving the overall S/N of the situation.

To sort rare cells from a mixture requires "specific labeling" of these cells. Everyone hopes that the rare cells can be separated on the basis of a single parameter (e.g. immunofluorescence using a highly specific monoclonal antibody). However, this is not usually the case. Even the most highly specific monoclonal antibodies usually are insufficient for unequivocal identification of the rare cells. Quite often, the number of false-positives is several times that of the number of true-positives. However, if multiple labels are used the situation

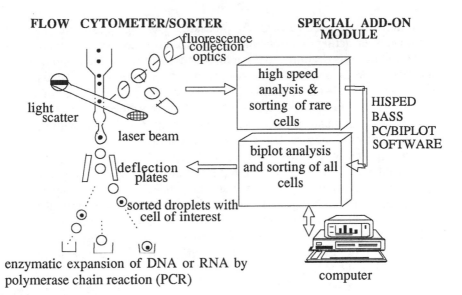

Figure 1: Special high-speed (HISPED) and principal component/biplot analysis and sorting system (BASS) modules shown on the right can be linked to existing commercial and home-built flow cytometer/cell sorters shown on the left. The DNA or RNA from rare sorted cells can be enzymatically amplified by polymerase chain reaction (PCR) to yield preparative amounts of DNA or RNA for Southern or Northern blotting analyses.

is usually improved. Each label, while having some amount of non-specificity, adds new information. More importantly, true-positive cells differ from the false-positive cells when they are labeled according to combined parameters. Properly chosen combinations of specific labels correlate with each other with a positive correlation on the true-positive cells; but non-specific labels on false-positive cells have little or no correlation with one another. This fact enabled us to use principal component/biplot sorting as described later in this chapter to separate true-positives from false-positives.

A need for high speed signal processing electronics

In the studies described in Cupp et al. (2), early experiments without the present high-speed system required nearly 6 hours to run a positive sample and another 6 hours to run the control sample. Use of special high-speed circuitry which allowed analysis at rates in excess of 100,000 cells/sec reduced the time per sample to approximately 15 minutes. In conventional cell sorters, the total instrument dead-time for processing each cell is on the order of 50-100 microseconds. These deadtimes cause most cells to be missed by the system when more than 20,000 cells/sec are processed. Our high-speed pre-processing circuitry, when acting as a front end-filter to a conventional flow cytometer/cell sorter, reduces this deadtime to 2 microseconds. This deadtime could be further improved by using one or more secondary buffering stages in the signal processing electronics. Few cells are missed by the system at rates in excess of 100,000 cells/sec. For example, at 150,000 cells/sec queuing theory based on random, Poisson arrival statistics predicts that only 1.9 percent of the cells will be missed due to instrument deadtime. Our experimental data agree well with that predicted by queuing theory, provided that the cells are not damaged and sticky and provided that the viscosity of the sample stream is equivalent to that of 0.25% bovine serum albumin in phosphate buffered saline. The actual deadtime of the system may vary to be slightly more than this depending on the size of the cells and the laser beam width, but is typically less than 3 microseconds for all applications to date. The high-speed system attempts to classify cells as "positive/not sure", or "negative". Only "positive/not-sure" cells are passed on to the analog-to-digital converters to be digitized.

For the application to Rh-incompatibility described in Cupp et al. (2), we were able to find fetal Rh^+ cells in maternal blood at frequencies as low as 10^{-5}. However, for other applications such as the isolation of fetal nucleated cells for subsequent genetic analysis the false-positive background from the maternal cells did not permit successful isolation of fetal cells on the basis of a single parameter. In fact, even multiparameter high-speed analysis of these fetal cells did not permit their successful isolation at very high purity. This led us to develop a second-stage system which looks at the correlations between multiple parameters on the "positives/not-sure" cells, not just their signal intensities as is done by conventional flow cytometry.

This second-stage processing unit BASS (Biplot Acquisition and Sorting System) (**Figure 2**) then attempts to correctly classify the "not-sure" cells into "true positives" and "false positives" so that only "true positives" are sorted. The joining of the high-speed and BASS systems is shown in more detail in **Figure 3**.

Sorting speed versus purity

Sorting of rare cells is limited by the number of droplets/sec. The problem is a straight-forward application of queuing theory. If cells arrive randomly, they are distributed among the subsequently-formed droplets according to a Poisson distribution. At typical sorting rates of 2000 cells/sec, the probability of two cells occurring inside the same droplet ("coincidence") is negligible. However, at higher rates the coincidence rate sharply increases. To sort cells of high purity (95%) at rates of 5000 cells/sec or greater requires "anti-coincidence" circuitry which rejects all cells which are close enough to be sorted in the same droplet or droplets (there may be more than one droplet in each sorting unit). At very high sorting rates (e.g. 100,000 cells/sec), there will be multiple cells in each sorted

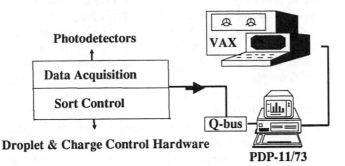

Figure 2: The BASS (Biplot Acquisition and Sorting System) module provides for both real-time acquisition and transformation of data in principal component space so that subpopulations of cells as seen in projections of higher dimensional space can be visualized by human observers. Biplot analyses reveal "true-positive" and "false-positive" cell subpopulations so that the "true-positives" rare cells can be sorted at very high purity.

High-Speed Multiparameter Rare Cell Analysis and Sorting

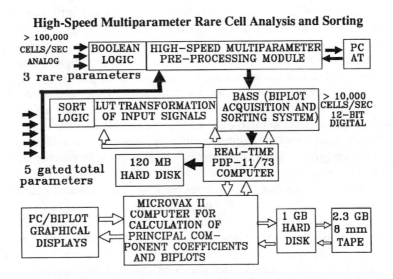

Figure 3: The high-speed module acts as a front-end signal processing filter allowing conventional flow cytometer/cell sorters to operate at rates in excess of 100,000 cells/sec. Only signals from "positives/not-sure" cells are passed along to the second stage principal component/biplot module which attempts to classify the "not sure" cells into "true-positives" and "false-positives". High speed lookup tables (LUT's) are used to transform input signals in real time to permit sorting of cells from principal component space.

droplet. While this is generally undesirable, the actual enrichment factor by sorting is greater at high sorting rates than at lower rates for the case of rare (0.1%) cell subpopulations. The enrichment factor for cell separation of a 10% cell subpopulation to become 95% pure is less than 10. However, for a high-speed separation of a 0.01% cell subpopulation at 120,000 cells/sec with a droplet rate of 30,000 droplets/sec, the average number of cells in each sorted droplet is 4. Importantly, one (and very rarely more than one) of these four cells will be a positive cell from the .01% rare cell subpopulation. Since the original frequency of rare cells is 0.01% and the final frequency is very slightly more than 25%, the enrichment factor is more than 2,500. Similarly, for separation of a 0.001% cell subpopulation, the enrichment factor is more than 25,000. Part of the BASS system involves "flexible sorting" strategies designed to allow the experimenter to select a yield vs. purity trade-off strategy most suitable for the goals of the experiment (U.S. patent pending).

Visualization of multidimensional rare-cell signals using principal components and biplots

A problem in visualizing multiparameter flow cytometry data is that humans can visualize only two dimensional data. Visual tricks can be used to create the illusion of depth to view three dimensions. For example, three-dimensional data can be shown using hidden-line techniques with solid iso-contours (7). Stereo-pairs can be used to allow stereoscopic viewing of three-dimensional data. New autostereoscopic displays can be used to allow direct viewing of three-dimensional data on a personal computer (8). One problem with these approaches is that as the data is spread in three-dimensions, the number of cells in each voxel (three-dimensional bin of the display) rapidly approaches zero unless very large data sets are analyzed. The problem becomes even more severe as the number of dimensions increases.

Dimensions higher than three cannot be directly observed. Unfortunately, successful isolation of rare cell subpopulations frequently requires four or more parameters to be measured on each cell. Four dimensional data are frequently viewed by looking at all non-redundant bivariate combinations of the data. For example, four dimensional data of flow cytometric parameters A, B, C and D would have 6 possible non-redundant bivariate views of the data: A vs B, A vs C, A vs D, B vs C, B vs D, and C vs D (where B vs A, etc. are considered redundant displays). For six-parameter data, there are 15 non-redundant bivariate displays. For eight-parameter data there are 28 non-redundant bivariate displays, and so on. Not only is it difficult to ask humans to attempt to reconstruct higher dimensional data from these visual "slices" of the data, but all of the slices do not necessarily portray the true data since the data will be oriented arbitrarily in higher dimensional space. Such slices will tend to project subpopulations down on top of each other. Clearly there is a need to find a way of visualizing multidimensional data in a single or a few optimal displays.

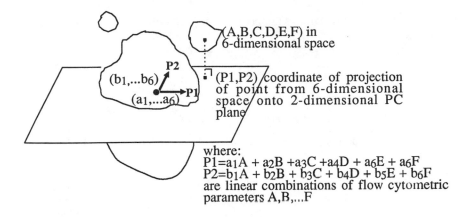

Figure 4: Principal component analysis projects higher dimensional data down onto a two-dimensional hyperplane which maximizes the variance of the data. In this way clusters of higher dimensional data corresponding to cell subpopulations can be visualized by human observers.

One of the ways higher dimensional data can be visualized is through the use of principal component analysis, a special case of the more general "projection pursuit analysis" (9). Use of the first two principal components chooses a coordinate system that yields a two dimensional graphical representation of multidimensional data in a way that shows the greatest variance in the data (**Figure 4**). However, this means that the first two principal components will tend to be determined by larger cell subpopulations. Other cell subpopulations, particularly smaller ones and very rare ones, will not be well represented in this choice of principal components or projection unless they lie far from the centroid of the data. Principal component analysis should not be used in "brute force" fashion. Careful application of principal component analysis can be used to optimize representation of particular cell subpopulations. The first step is to consider the first two principal components only a "rough first cut" through the multidimensional data-space. The larger cell subpopulations which may not be of interest can then be removed from the data set by gating and reprocessing the data on the basis of principal components and/or raw parameters. New principal components can then be calculated on this subset of correlated, reprocessed listmode data. Such gating, reprocessing and reprojection of the data onto a new principal component plane may reveal other cell subpopulations hidden by the original principal component projections which will be chosen to maximize the variance of the major (rather than the minor) cell subpopulations.

Principal components are linear combinations of the original flow cytometric parameters. The first principal component P_1 is that linear combination of flow cytometric parameters that gives the maximum spread (variance) of the data

Correlation Matrix   **Biplot Column Markers**

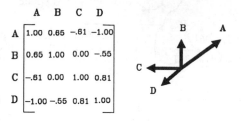

Figure 5: A biplot is a way of graphically representing the correlation matrix and projecting it down into an observable two-dimensional plane (e.g. the principal component plane).

points. The second principal component P_2 is the linear combination which gives the next greatest spread of data, and so on for other principal components. For example,

$P_1 = a_1A + a_2B + a_3C + a_4D + a_5E + a_6F$
$P_2 = b_1A + b_2B + b_3C + b_4D + b_5E + b_6F$
where $a_1, ... b_1, ...$ represent the coefficients
and A, B, ... F represent flow cytometric parameters
such as green fluorescence, red fluorescence, light
scatter, etc.

In a biplot (*9,10*) the principal component parameters are replaced by "row markers" which are proportional to principal components. The "row markers" (for each cell) are plotted together with "column marker" arrows each of which represents a flow cytometric parameter (e.g. green fluorescence). The length of each arrow is proportional to the standard deviation of the data represented by that parameter. The cosine of the angle between any two arrows is proportional to the correlation between the parameters (see **Figure 5**). Both the length of the arrows and the angles between them may, however, be distorted by the process of projection onto a principal component or other plane.

Using biplots to distinguish between true-positive and false-positive cell subpopulations

Even with the best staining of rare cells, there will always be background and noise leading to "false-positive" cell subpopulations. One of the useful features

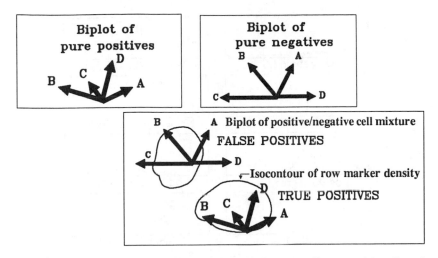

Figure 6 : Biplots can be used to distinguish between "true-positives" and "false-positives" for high-resolution sorting of rare cell subpopulations. Each subpopulation of cells will have a pattern of biplot vectors which is unique for that subpopulation. "False-positives" will have a different biplot pattern than "true positives" because biological "noise" tends to "de-correlate" the flow cytometric variables.

of biplots we are exploring is their ability to distinguish between "true-positive" and "false-positive" cell subpopulations as visualized by biplot analyses (**Figure 6**). By examining the pattern of the biplot vectors we can often distinguish between the "true-positive" and "false-positive" cell subpopulations. Such information can be used to obtain highly purified rare cells of interest.

An example of high-speed, principal component/biplot analysis of a rare cell subpopulation

To illustrate all of the preceding discussion of high-speed, principal component/biplot analysis we show in **Figure 7** actual data obtained on a rare (0.01 percent) population of peripheral blood mononuclear cells labeled by fluorescence in-situ hybridization (FISH) (*12*) analyzed through the high-speed system (HISPED) and biplot acquisition and sorting system (BASS) at a rate of 50,000 total cells/sec. The cells were also counterstained with propidium iodide (PI) (at less than saturating dye concentrations to prevent overwhelming of the FITC signal) and gated on the G_0/G_1 peak to eliminate debris, damaged nuclei, and cell doublets. High-speed flow cytometry was performed on a mixture of FISH-stained male human lymphocytes (0.01 percent) within a larger population of human whole blood. A biotinylated human repeat Y-chromosome specific probe (Amersham, Int., England) was labeled with FITC conjugated Extravidin (Sigma

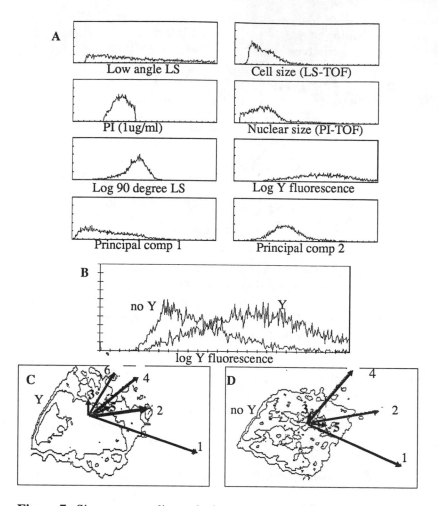

Figure 7: Six-parameter listmode data was obtained by high-speed analysis using the HISPED and biplot acquisition and sorting system (BASS) described previously. A rare (.01 percent) population of male human lymphocytes was labeled by suspension fluorescence in-situ hybridization (FISH) with Y-chromosome specific probe (cf. main text for details)Figure 7A. Principal components were calculated on a cell-by-cell basis and subsequently included as correlated listmode parameters 7 and 8. Comparisons of this sample (Y) were made with that of a control sample (no Y) in Figure 7B. Biplots for the two samples were calculated in Figures 7C and **7D**. Isocontour plots were made of the row markers (cells)and column markers (the 6 arrows corresponding to the 6 flow cytometric parameters) (cf.main text for details). (LS = light scatter, TOF = time-of-flight).

Chemical Company, St. Louis) and amplified once using biotinylated anti-avidin antibody. Figure 7A shows the 6-parameter correlated listmode data that was obtained on the rare cells passed through the HISPED system for subsequent digitization and analysis by principal component analysis. The first two principal components were calculated on a cell-by-cell basis as previously described and subsequently added as correlated listmode parameters 7 and 8. Figure 7B shows an overlay of the log Y-probe fluorescence of histogram 6 in Figure 7A with that of the control sample which has been treated identically except that the Y-probe itself has been singularly omitted. Figures 7C and 7D show the biplots of the Y-probe and no Y-probe samples respectively. The biplots are projected onto common planes through forced projection from their original (and different) principal component planes. Each cell constitutes a row-marker of the biplot, while each of the 6 flow cytometric parameters constitutes a column marker of the biplot (as shown by the arrows). Rather than plotting each row marker, cell isocontours have been drawn through a density plot of the row markers as is traditionally done with bivariate isocontour plots of conventional flow cytometric parameters. Alternatively, lookup tables of principal component values can be downloaded into hardware allowing multiple gating and reprojections in real-time to permit cell sorting. Special "flexible sorting" anticoincidence algorithms implemented in hardware (U.S. patent pending) can be used to sort highly purified rare cells for subsequent analysis by polymerase chain reaction (PCR) and other methods.

PCR expansion of specific DNA or RNA from rare sorted cells

The advent of new technologies outside the field of cell separation can have major impact on the practicality of some cell separation techniques. Until recently cell sorting has been confined mainly to the cloning of single or small numbers of cells for growth in tissue culture or for subsequent assays which require only small numbers of cells. Unlike some of the other cell separation methods described in this book, cell sorting can not sort preparative amounts of cellular material for traditional biochemical analysis. However, the invention of polymerase chain reaction (PCR) changes this picture drastically in terms of DNA or RNA which can be enzymatically expanded rapidly in vitro (*13*). Thus, a very small number of sorted cells (even a single cell) can have portions of its DNA or RNA enzymatically expanded to produce the equivalent amount of this DNA or RNA as if one had sorted 10^6 to 10^9 cells. Subsequently, standard molecular biology techniques can be used on the PCR expanded DNA or RNA from small numbers or even a single rare sorted cell. The trick becomes to accurately sort the correct cell, i.e. to perform high-resolution sorting of rare cells, the theme of this chapter. Such high-resolution sorting of rare cells when combined with PCR will make possible many applications in biology and medicine.

Some applications of high-speed, principal component/biplot sorting

In **Figure 8** we show three applications of high-speed, principal component/biplot sorting ongoing in our laboratory and the laboratories of collaborators. Our own research interest is the high-resolution sorting of rare human fetal cells from maternal blood for a risk-free, non-invasive prenatal diagnosis of birth defects. Human fetal cells are present in maternal blood at frequencies

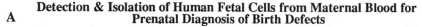

A   **Detection & Isolation of Human Fetal Cells from Maternal Blood for Prenatal Diagnosis of Birth Defects**

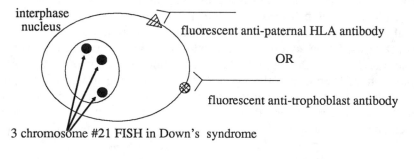

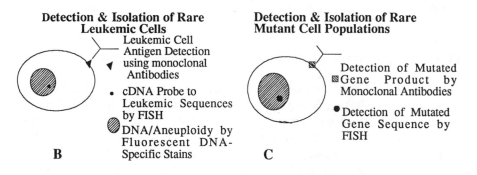

B   C

Figure 8: Three applications of high-speed, principal component/biplot sorting are shown. Fluorescence in-situ hybridization(FISH) and/or monoclonal antibodies against paternal histocompatibility (HLA) or trophoblastic antigens can be used in a risk-free assay to detect and isolate human fetal cells directly from maternal blood. (Figure **8A**) Rare leukemic cells can likewise be labeled either by FISH or by monoclonal antibodies against leukemic cell antigens for subsequent detection and isolation by high-speed cell sorting.(Figure **8B**) Rare mutant cell subpopulations can be identified and isolated on the basis of mutated gene sequences or products. (Figure **8C**)

between 10^{-7} and 10^{-4}. Fetal cells can be sorted from maternal blood by the techniques just described. Sorted fetal cells can have portions of their DNA enzymatically amplified by polymerase chain reaction (PCR) for further genetic analysis by Southern blotting (*14*). A second application involves separation of rare metastatic cancer cells from peripheral blood or from bone marrow. Detection of these rare tumor cells can be extremely useful for monitoring patients in "remission" who still have tumor cells in their bodies but at frequencies undetectable by other methods. The body's immune system will deal effectively with some types of tumor cells if the tumor load does not overwhelm the body's immune system. A third application is the detection and isolation of rare mutant cell subpopulations for detection of somatic gene mutations, both spontaneous and induced by chemicals, radiation, or other environmental factors. Accurate assessment of somatic mutations for genetic toxicology and biological dosimetry could serve as monitors of the environment and for monitoring hazards to workers on this planet and for astronauts in space.

Summary

In summary, rare cells can be analyzed and sorted on the basis of multiple parameters at high-speeds. False-positives can be reduced using principal component/biplot analysis. PC/biplot analysis can be implemented in hardware to permit real time gating, reprojections, and sorting necessary to reveal rare cells hidden by more major cell subpopulations.

Literature Cited

1. Langlois, R.G.; Bigbee, W.L.; Kyoizumi, S.; Nakamura, N.; Akiyama, M.; Jensen, R.H. *Science* **1987**, *236*: 445-448.
2. Cupp, J.E.; Leary, J.F.; Cerniachiari, E.; Wood, J.C.S.; Doherty, R.A. *Cytometry* **1984**, *5*: 138-144.
3. Herzenberg, L., Bianchi, D., Schroder, J., Cann, H., Iverson, M. *Proc. Nat. Acad. Sci.* **1979**, *76*:1453-1455.
4. Ryan, D.H.; Mitchell, S.J.; Hennessy, L.A.; Bauer, K.D.; Horan, P.K.; Cohen H.J. *J. Immunol. Methods* **1984**, *74*: 115-128.
5. Visser, J.W.M.; Martens, A.C.M.; Hagenbeek, A. *Ann NY Acad Sci* **1986**, 468: 268-275.
6. Peters, D.; Branscomb, E; Dean, P; Merrill, T.; Pinkel, D; VanDilla, M; Gray, J.W. *Cytometry* **1985**,*6*: 290-301.
7. Valet, G. *In Flow Cytometry IV*, Laerum, O.D., Lindmo, T., Thorud, E. (Eds.), Universitetsforlaget, Bergen, Norway. **1980**, pp. 125
8. Leary et al. Cinical Cytometry Meeting, Charleston, SC., Sept.12-15, **1990**.
9. Huber, P. J., *Annals of Statistics* **1985**, *13*:435-525.
10. Gabriel, K.R. *Biometrika* **1971**, *58*: 453-467.
11. Gabriel, K.R.; Odoroff, C.L. In *Data Analysis and Informatics III*, Diday

et.al., Editors, Elsevier Science Publishers, B.N. (North Holland), **1984**, pp 23-30.

12. Trask, B., Van den Engh, G., Landegent, J., in de Wal, N.J., Van der Ploeg, *M. Science* **1985**, *230*:1401-1403.

13. Mullis, K., Faloona, F., Scharf, S., Saiki, R., Horn, G., Erlich, H., In Cold *Spring Harbor Symposia on Quantitative Biology*, **1986** vol. *51*, pp. 263-273.

14. Southern, E.M. *J. Mol. Biol.* **1975**, *98*: 503-517.

RECEIVED March 15, 1991

Chapter 3

Rare-Earth Chelates as Fluorescent Markers in Cell Separation and Analysis

R. C. Leif[1] and L. M. Vallarino[2]

[1]Coulter Electronics, Inc., 690 West 20th Street, Hialeah, FL 33010
[2]Department of Chemistry, Virginia Commonwealth University, Richmond, VA 23284

The synthesis of a novel series of functionalized macrocyclic complexes of the lanthanide(III) ions is reported. The Eu(III) complexes possess a set of properties (water solubility, inertness to metal release, ligand-sensitized luminescence, reactive peripheral functionalities) that make them suitable as luminescent markers for bio-substrates. Currently employed organic fluorophores give efficient signal production, but this is accompanied by interference from background fluorescence and, if multiple fluorophores are used, from spectral overlap. The long lifetimes and narrow-band emissions of the luminescent lanthanide complexes will minimize background interference; however, long lifetimes will also result in a significantly reduced signal for flow cytometry or cell sorting.

Cell separation and cell analysis are mutually dependent branches of cytophysics. The quality of a cell separation can be ascertained by analyzing the individual fractions to determine their cellular composition. In turn, the accuracy of a cell analysis procedure can be documented most easily for a cell population that has been enriched in a specific fraction by an effective separation procedure. In the past, it was sufficient to report the results of a cell separation in terms of morphology (1-3). Recently, however, the development of monoclonal antibodies and of recombinant DNA techniques has greatly increased the capability to analyze cells for specific antigens and even to observe single genes (4) thus permitting the use of these advanced detection procedures.

0097–6156/91/0464–0041$06.00/0

The Advantages of Narrow-Line Emitting Fluorophores

Current fluorescent histochemical techniques, whether based on flow analysis or on image cytometry, are limited by the chemical nature of the fluorescent markers. Traditional fluorophores are large, rigid organic molecules in which both the ground and the excited electronic states are comprised of a multitude of closely spaced vibrational levels. In these molecules the electrons involved in light absorption are outer shell (valence) electrons; their energies are affected by atomic vibrations. The photons absorbed therefore have an energy distribution which corresponds to the many possible values of the energy difference between the lowest vibrational level(s) of the ground electronic state and various vibrational levels of the excited electronic state. Almost immediately after light excitation, these molecules relax to the lowest vibrational level of the first excited state. When fluorescence emission occurs, the energy distribution of the emitted photons corresponds to the many possible values of the energy difference between this lowest vibrational level of the first excited state and various vibrational levels of the ground state. At room temperature, therefore, both the excitation and the emission spectra are broad (5). With traditional organic fluorophores, overlap between the fluorescence of the marker and the broad autofluorescence of the cellular substrate represents a major limitation in the sensitivity of the optical measurements (6-7). The broadness of the emission-excitation spectra also severely reduces the number of fluorophores that may be measured simultaneously (8). The emissions of fluorescein and rhodamine, for example, overlap to such an extent that in flow microfluorometers and other analytical instruments special electronic circuitry must be employed to separate the individual signals (9). Although this type of dual measurement has been considerably facilitated by the substitution of rhodamine with phycoerythrin (10), which absorbs at lower wavelengths, the limitation remains.

The preceding considerations led Leif et al. to propose the use of rare-earth chelates as luminescent reporter molecules (11). In these compounds, the luminescence results from transitions within the 4f electron manifold of the metal ion. Since the inner shell 4f electrons are non-bonding and are effectively shielded from the environment by the outer electron shells, these emissions are free from vibrational broadening (12).

Figure 1 shows the most intense emissions of Eu(III) and Tb(III) superimposed for comparison on the emissions of fluorescein and rhodamine; clearly, these lanthanide emissions do not overlap.

The Question of Long Lifetimes. Another distinctive feature of the luminescent lanthanide chelates is their long excited-state lifetime, usually more than 300 μs (13). Since the lifetimes of natural as well as conventional synthetic fluorophores are usually of the order of 10 ns

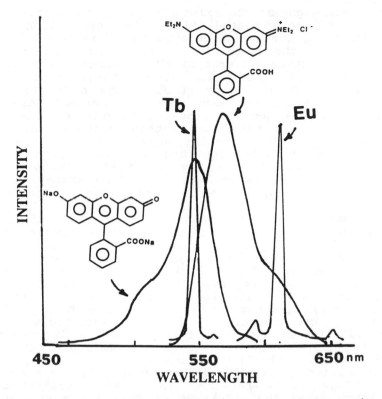

Figure 1.   Emission profiles of fluorescein, rhodamine, and of Eu(III) and Tb(III) nitrates in aqueous solution. (Emission intensities on arbitrary scales.)

(5), the lanthanide luminescence may be completely freed from background interference by short-pulse excitation of the sample, followed by slightly delayed signal detection (delay, ca. 1 μs). These time-gated measurements based on the long-lived luminescence of the lanthanide marker are advantageous in automated immunoassays (14) and would also be of value in the microscopic examination of cells as well as in flow cytometry or flow cell sorting. In the latter case, however, previous analysis (15) has shown that the signal intensity will also be diminished. Even in the favorable situation of a 10 μs excitation pulse and a lanthanide chelate with a relatively short lifetime of 38 μs, the percentage of light emitted within the orifice will be only 34%. This includes the light signal collected while the cell is still being illuminated. True dark field observation with a conventional continuous wave laser would decrease this to less than 25%. With long-lived lanthanide fluorophores as markers, slowing the flow rate of the sample would significantly increase the efficiency of signal collection.

A pulsed light source would also provide a significant increase in collected signal (16). Pulsed excitation would reduce the time, during the observation of the cell in the orifice, when the detection system must be either electronically gated off or optically obscured in order to achieve dark field illumination. Two suitable pulsed light sources are short arc xenon strobe lamps and noble gas lasers. Both argon and neon lase with markedly greater efficiency when pulsed and produce ultraviolet emissions suitable for the excitation of lanthanide complexes.

The luminescence lifetime of europium chelates is longer than the transit time through a transducer (10-30 μs). Therefore, the question may be raised of whether such chelates, when used as markers in a flow system, would produce a significantly lower signal than conventional short-lived fluorophores. Mathies and Stryer (17) have stated that, under saturating laser excitation, a fluorophore is in principle capable of being excited once for each lifetime; for fluorescein, this would mean approximately 200 times per μs. The excitation conditions assumed by Mathies and Stryer in their calculations were different from those employed in a modern flow cytometer such as the EPICS Elite (EPICS is a trademark of the Coulter Corp.). These authors assumed a beam with a circular 4 μm cross-section at the $1/e^2$ points and 65 mW of 488 nm radiation energy. With these parameters, Mathies and Stryer calculated that one fluorescein molecule could be excited approximately 97 times per μs.

The situation is quite different in a flow cytometer. At the flow rates required to obtain statistically significant data, the thickness of the sample stream is sufficient to permit some lateral position dispersion of the cells in the flow chamber. To provide consistent laser excitation of cells with imperfect trajectory, the gaussian laser beam is broadened in the direction perpendicular to the flow. The size of the laser beam at the $1/e^2$ points in an EPICS Elite is approximately 35X75 μm^2, an area 164 times greater than that assumed in the calculation by Mathies and Stryer. The air-cooled argon ion laser in the EPICS Elite is usually set to produce 15 mW of 488 nm radiation. Correction for both the increased area of the beam cross-section and the decreased radiation intensity results in a 711 fold decrease in excitation energy to the sample. Under the best conditions, this is equivalent to 0.129 exciting photons per molecule per μs. With the same geometric parameters, a 50 mW water-cooled UV Argon laser would excite 0.32 molecules per μs and a 9 mW UV He-Cd laser would excite 0.05 molecules per μs. If a combined DNA and Eu-chelate analysis were to be performed, the beam would be broadened. Multimode He-Cd lasers produce larger beams than gaussian lasers. In a ten μs excitation interval with the 50 mw UV Argon laser, approximately (10x0.32)=3.2 photons would be available to excite each chelate. This number could be increased by employing a high powered continuous or pulsed laser; however, such sources are not as yet available in clinical flow cytometers.

The above calculations, based on the cited work of Mathies and Stryer, indicate that the inability of the Eu-macrocycle to undergo multiple cycles of excitation and emission during the time of flow detection may result in an approximate 3-fold decrease in useful signal. However, preliminary studies indicate that luminescent Eu-macrocycles undergo less concentration quenching than traditional organic fluorophores. Thus, this limitation may in principle be overcome by greater loading of the luminescent marker on the substrate.

Work by Other Investigators

The usefulness of lanthanide complexes as biological markers has been recognized by other workers. Soini and Hemmila (18) and later Soini and Lovgren (14) have described an analytical procedure involving the Eu(III) complex of the polyamino-carboxylate ligand DTPA (19) (Fig. 2(a)), which is non-luminescent in aqueous solution. In this procedure, the Eu-DTPA chelate was decomposed with acid and the solubilized Eu(III) was complexed with a ß-diketonate in a micellar phase (20). These dissociation-combination steps complicated the analytical procedure and increased the volume of solution to be monitored for Eu(III) emission, with consequent decrease in optical sensitivity. Since this protocol involved separation of the fluorophore from the specific bio-substrate, the technique was unsuitable for immuno-luminescence and other measurements on single cells or particles by either flow cytometry or microscopy.

Another example of luminescence analysis based on a Eu(III) complex has been reported by Evangelista et al. (21). In this case the ligand was the dinegative ion of 4,7-bis(chlorosulfenyl)-1,10-phenanthroline-2,9-dicarboxylic acid, BCPDA (Fig. 2(b)), which formed a luminescent chelate with Eu(III). The luminescence, however, was not detectable in aqueous solution at the concentrations of the analytes of interest and the sample had to be dried prior to measurement. Furthermore, the intrinsic luminescence of the Eu-BCPDA complex was rather low; in order to measure analytes present in minor quantities, it was necessary to increase sensitivity by binding the chelate to thyroglobulin (22) in a multilayer system. Soini et al. (23) have observed luminescence from a chelate attached by antibody to a cell, but do not report the chemical structure of the chelate. Beverloo et al. (24) have reported the use of microcrystals of the phosphor yttrium(III) oxysulfide doped with europium(III). Cells specifically labeled with this phosphor could be detected in a time- gated mode even in the presence of the very bright, red emitting DNA stain, ethidium bromide. A disadvantage of these 50-500 nm phosphor particles is that they must be coated to prevent agglutination with a

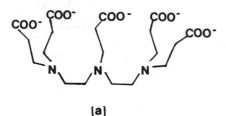

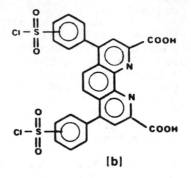

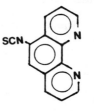

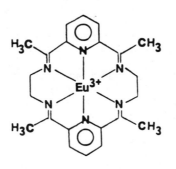

Figure 2. Schematic formulas of: (a) Diethylenetriamine-pentacetate, DTPA; (b) 4,7-bis(chlorosulfenyl)-1,10-phenanthroline-2,9-dicarboxylate, BCPA; (c) A generic ß-diketonate (R and R' may be equal or different); (d) 5-isothiocyanato-1,10-phenanthroline; (e) Eu-L^1 complex, where L^1 represent the macrocyclic ligand C$_{22}$2H$_{26}$N$_6$.

polycarboxylic acid, which may then be displaced by negatively charged ions such as phosphate, citrate, or ionized ethylenediamine tetracarboxylic acid (EDTA). The negative charge of these coated phosphors also causes nonspecific binding to slide coating agents such as poly-l-lysine or bovine serum albumin.

Results and Discussion

This section describes our systematic efforts to design and synthesize water-soluble, chemically stable, luminescent complexes of the Eu(III) and Tb(III) ions, equipped with peripheral functional groups for coupling to a bio-substrate. Details of the experimental methods and results are to be found in the references cited.

Our search for lanthanide derivatives suitable as luminescent markers initially focused (11,15) on the tris-chelate complexes of ß-diketonates (Fig. 2(c)). These compounds possessed three important favorable features. First, the tris-ß-diketonates of both Eu(III) and Tb(III) were known to exhibit ligand-sensitized luminescence, with emission intensity depending largely on the efficiency of the energy transfer from an excited triplet state of the ß-diketonate ligand to the emission level(s) of the metal ion (25). Second, the lanthanide tris-ß-diketonates were thermodynamically stable, with cumulative stability constants, β_3, ranging from 18 to 21 log units (26-27). Third, the tris-ß-diketonates contained several coordinated water molecules that could be replaced by uncharged N-donor heterocyclic ligands (28) to give complexes having decreased vibrational quenching and hence increased luminescence. It was reasonable to conclude that, if the N-donor heterocycle were a chelating ligand previously coupled to a bio-substrate, this substitution reaction could provide a straightforward way of attaching the luminescent complex to the intended target.

To explore this possibility, we developed the synthesis of the previously unknown bifunctional chelate, 5-isothiocyanato-1,10-phenanthroline (Fig. 2(d)) (29). This was coupled to a model protein (BSA and γ-globulin) via a thiourea linkage, following the procedure established for fluorescein isothiocyanate. We then proceeded to attach a preformed Eu(III) or Tb(III) tris-ß-diketonate to the protein-coupled phenanthroline. However, this last step met with failure, in that the highly luminescent protein-lanthanide conjugate initially formed dissociated upon washing or dialysis (29). The phenanthroline moiety remained covalently attached to the protein, while the lanthanide complex broke down and luminescence was lost. Obviously, the thermodynamic stability of these complexes, however high, was not sufficient to prevent their dissociation in a very dilute, buffered aqueous solution.

These first unsuccessful attempts emphasized an important generalization, namely, that the time frame of ligand exchange and metal exchange kinetics becomes a major consideration for complexes to be used as probes in biological systems. In the very dilute aqueous or aqueous-organic media required for such systems, often involving contact with potentially competing ligands, even ordinarily "stable" metal complexes, if labile, may dissociate. When such dissociation occurs, the value of the probe is diminished or lost. The task, therefore, was to design and synthesize a new class of Eu(III) and Tb(III) complexes that would retain their identity in dilute aqueous solution through kinetic inertness rather than through thermodynamic stability alone.

To achieve this objective, we decided to take advantage of the so-called macrocyclic effect. Examples of metal-macrocyclic complexes that remain undissociated in solution, even though containing inherently labile metal ions, abound both in nature and in the synthetic chemical literature (30). The vast majority of these macrocyclic complexes contain a "small" metal ion bound inside the four-donor-atom cavity of an organic ligand, a striking example being the Mg(II) of chlorophyll. A limited number of inert complexes of larger metal ions with five-and six-donor-atom macrocyclic ligands have also been reported. Among these, the most relevant were the complexes of La(III) and Ce(III) with the six-nitrogen ligand L^1 (Fig. 2(e)), obtained by Backer-Dirks et al. (31) from the metal-templated cyclic Schiff-base condensation of 1,2-diaminoethane and 2,6-diacetylpyridine. Although these authors had been unable to similarly synthesize the corresponding complexes of other lanthanides, an appropriate change in experimental procedure, together with use of the lanthanide acetates instead of the nitrates as templates, allowed us to obtain the Eu(III) and Tb(III) complexes in high yields and excellent purity (32).

These complexes were white crystalline solids, thermally very stable and soluble in water as well as common organic solvents. They were unique among the derivatives of the lanthanide(III) ions in that the metal-macrocycle entities remained undissociated in dilute aqueous solution, even under conditions that would result in rapid decomposition of most other lanthanide complexes. In contrast, the exocyclic ligands, whether anions or solvent molecules, were labile and readily exchangeable. For example, prolonged contact of the Eu-L^1 acetate complex with a ten-fold excess of hydrochloric acid in methanol resulted in replacement of the acetates by chlorides without degradation of the Eu-L^1 entity. The behavior of the M-L^1 complexes in solution is consistent with their structure as established by single-crystal X-ray analysis (Benetollo, F.; Bombieri, G.; Fonda, K.K.; Polo, A.; Vallarino, L.M. Polyhedron, in press).

In $EuL^1(CH_3COO)_2Cl \cdot 4H_2O$ (Fig. 3), the Eu ion is bound to the six N atoms of the L^1 macrocycle in a nearly planar arrangement, with Eu-N internuclear distances corresponding to the sum of the individual radii. The two bidentate chelating acetate ligands occupy axial positions on opposite sides of the Eu-macrocyle unit, the organic portion of which is slightly folded in a "butterfly configuration" that minimizes steric strain.

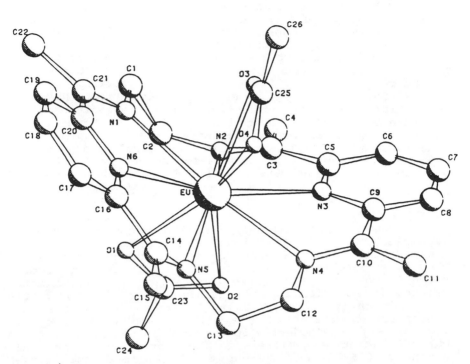

Figure 3. Structure of the [EuL^1(CH$_3$COO)$_2$]Cl·4H$_2$O complex in the crystalline state, showing the atom labeling scheme. The oxygens of the clathrated water molecules are omitted for clarity.

The ease of exchange of the exocyclic anions proved to be a major asset in view of the potential use of these compounds as luminescent markers. The Eu-L^1 and Tb-L^1 acetate-chloride salts, initially obtained from the metal-templated synthesis, exhibited only a weak luminescence (*33–34*); however, substitution of the acetates by a variety of chelating carboxylates or ß-diketonates resulted in highly luminescent species. The most effective luminescence enhancers were mononegative ligands with rigid, π-bonded structures and hard donor atoms (O, aromatic N) (*35*). Figure 4 illustrates the increase in emission intensity observed upon addition of 2-furoic acid to the Eu-L^1 diacetate-chloride. The stoichiometric character of this effect is evident from the luminescence titration graph. Substituted macrocycle-enhancer complexes were also isolated and characterized as pure crystalline solids (*34–35*).

Having succeeded in synthesizing water-soluble complexes of Eu(III) and Tb(III) that could retain their identity in dilute solution and exhibit ligand-sensitized luminescence, our next goal was to introduce into the backbone of the macrocycle a functional group suitable for coupling to a bio-substrate. We chose the synthetic scheme shown in Fig. 5, which utilized a carbon-substituted 1,2-diaminoethane precursor. This synthesis had the advantage of introducing the coupling groups into the flexible $-CH_2-CH_2-$ side-chains, at a position and at a distance where it would be less likely to disturb either the conformation of the macrocyclic ligand or the electronic character of the ligand-metal bonding. There was no precedent in the literature for this kind of synthetic procedure. However, we anticipated that under appropriate conditions the template action of the metal ion, favoring the formation of the macrocyclic complex, would prevail over the competitive side-reactions arising from the functional group of the diamine precursor. This expectation proved to be true and lanthanide complexes were obtained in good yields for the three macrocyclic ligands L^2 (with functional group X = -CH_2-OH), L^3 (with X = -CH_2-C_6H_4-OH), and L^4 (with X = -CH_2-C_6H_4-NH_2).

These functionalized metal-macrocycles closely resembled their non-functionalized analogs. They were microcrystalline solids with excellent thermal stability. They were soluble in water as well as in common organic solvents, and they exhibited the same inertness to metal release in solution as well as the same lability of the exocyclic anions. Also, similar to their L^1 analogs, the lanthanide complexes of the L^2, L^3, and L^4 macrocyclic ligands were only modestly luminescent as the acetate salts but became intense emitters in the presence of a suitable enhancer (36-37.). Figure 6 illustrates the marked increase in emission intensity that occurs when the Eu-L^3 triacetate is treated with gradually increasing amounts of 4,4,4-trifluoro-1(2-thienyl)butane-1,3-dione. The shift in the excitation maximum of the macrocycle-enhancer system, relative to that of the original macrocycle acetate alone, clearly shows that the chelating enhancer not only provides better protection from vibrational quenching by the solvent but also acts as a radiation absorber and promotes effective energy transfer to the metal ion.

It should be noted that a disubstituted macrocycle such as L^2, L^3, or L^4 can exist as two regioisomers, as illustrated in Fig. 7. One isomer is the cis form, in which the two functional groups X occupy positions adjacent to the same pyridine ring and thus are "on the same side" relative to the metal center. The other isomer is the trans form, in which the two X groups occupy positions adjacent to different pyridine rings and thus are "on opposite sides" relative to the metal center. Furthermore, the carbon atoms of the diimine side-chains which carry the X groups are chiral; thus, additional isomers become possible if the diamine precursor is racemic. The presence of isomers most likely is responsible for the failure we

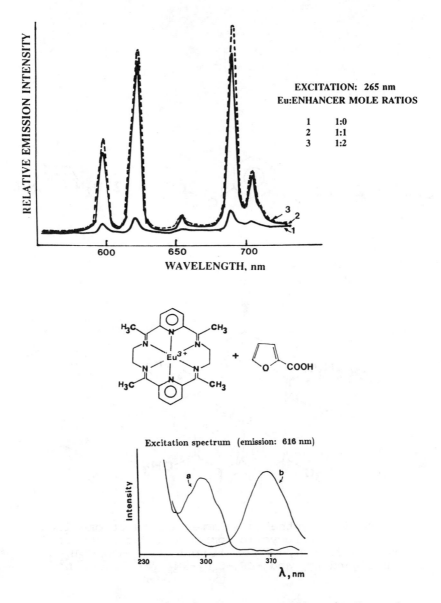

Figure 4. Luminescence "titration" of $[EuL^1(CH_3COO)_2]Cl \cdot 4H_2O$ with 2-furoic acid in methanol. Inset shows the excitation spectra of: (a) original complex and (b) complex with added enhancer (1:1 mole ratio).

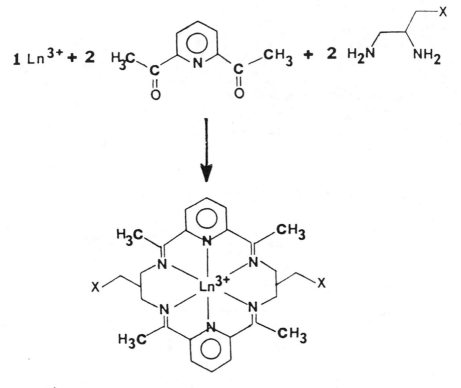

Figure 5.   Synthetic scheme for functionalized lanthanide macrocyclic complexes. The functional group X is $-CH_2-OH$ for ligand L^2, $-CH_2-C_6H_4-OH$ for ligand L^3, and $-CH_2-C_6H_4-NH_2$ for ligand L^4

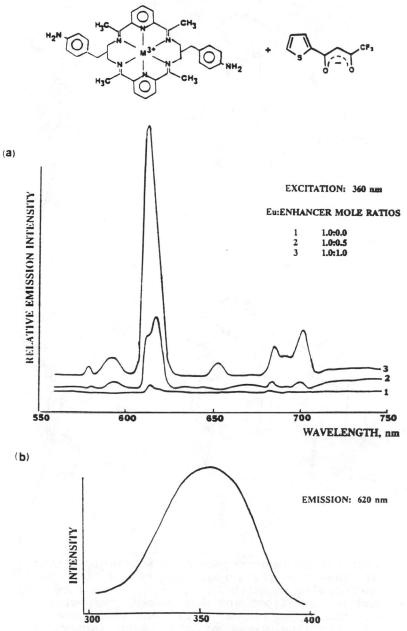

Figure 6.   Luminescence   "titration"   of
$EuL^3(CH_3COO)_3 \cdot nH_2O$ with   4,4,4-trifluoro-1(2-
thienyl)butane-1,3-dione   in   methanol: (a)
emission   spectra   of   solutions   containing
increasing   Eu   to   enhancer   mole   ratios,   (b)
excitation   spectrum   of   1:1   complex-enhancer
solution.

(a)

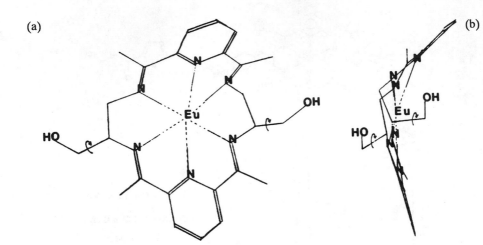

(b)

(c)

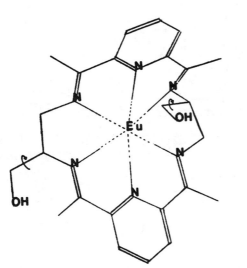

(d)

Figure 7. Computer-generated schematic formulas
of the two [S,S]-regioisomers of the Eu-L^2
macrocycle, in which X = -CH$_2$-OH. Structures (a)
and (b) are front and side views, respectively,
of the <u>cis</u> isomer; (c) and (d) are the
corresponding views of the <u>trans</u> isomer. For
these molecules, the functional groups of the <u>cis</u>
isomer occupy opposite "hemispheres" relative to
the plane of the macrocycle, whereas those of the
<u>trans</u> isomer occupy the same "hemisphere". The
carbon-attached hydrogen atoms of the macrocycle
and the exocyclic ligands are omitted for
clarity.

have so far encountered in obtaining single crystals of these functionalized macrocycles, suitable for X-ray analysis.

The peripheral -OH and $-NH_2$ functional groups of the macrocycles were found to exhibit their normal reactivity. For example, reaction of the metal macrocycle acetates with acetic anhydride, in the presence of 4-dimethylaminopyridine as catalyst, gave the corresponding ester or amide in good yields (38). No degradation of the metal-macrocycle entity occurred even under the drastic conditions required for this reaction and the resulting derivatives were soluble and inert to metal release. Figure 8 shows the infrared spectrum of the acetate ester of the M-L³ triacetate complex and illustrates the characteristic features of this type of compound.

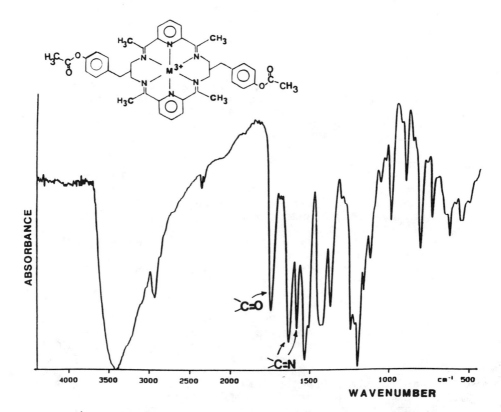

Figure 8. Infrared spectrum of the diacetylester of EuL³(CH₃COO)₃·nH₂O.   The peak at 1750 cm⁻¹ represents the stretching absorption of the carbonyl ester group.   The two peaks at 1634 and 1589 cm⁻¹ represent the C=N stretching absorptions of the pyridine and imine groups, respectively, and are a diagnostic feature of this kind of macrocycle.

Conclusions

The series of functionalized macrocyclic complexes of Eu(III) and Tb(III) reported in this paper fulfill the fundamental requirements of a luminescent marker for cytology and immunology. These complexes are soluble in water and water-compatible solvents; they do not release the metal ion even in the presence of acids, bases, or competing ligands; they can be made instantaneously luminescent in aqueous solution by the simple addition of an enhancer; finally, they contain primary hydroxy or amino functionalities that can be used for coupling to a desired bio-substrate. These functionalized macrocycles exhibit the narrow-emission, long-lived luminescence that is typical of traditional Eu(III) and Tb(III) chelates, and thus offer the same advantages in regard to sensitivity and lack of mutual or background interference. Future work will focus on the attachment of the functionalized macrocycles to bio-substrates and on the evaluation of the stability and luminescence of the resulting conjugates.

Acknowledgments

This work has been supported by Coulter Electronics, Hialeah, FL, by Virginia Commonwealth University, and by N.A.T.O. Bilateral Project No. 185-85. We wish to thank Dr. M. L. Cayer for editorial assistance and Mrs. M. Warren for her assistance in manuscript preparation.

Literature Cited

(1)   Leif, R.C.; Smith, S.; Warters, R.L.; Dunlap, L.A.; Leif, S.B. *J. Histochem. Cytochem.* **1975**, *23*, p. 378.
(2)   Hirsch, M.A.; Lipner, H.; Leif, R.C. *Cell Biophys.* **1979**, *1*, p. 93.
(3)   Pretlow II, T.G.; Pretlow, T.P. In *Cell Separation Methods and Selected Applications*; Pretlow II, T. G.; Pretlow, T.P., Eds.; Academic Press; San Diego, **1982**, Vol.1; Chapt.2.
(4)   Stewart, C.C. *International Conference on Analytical Cytology XIV,* **1990,** Abstr. 16.
(5)   Lakowicz,   J.R.   *Principles   of   Fluorescence Spectroscopy*; Plenum, N.Y., **1983**; p. 5.
(6)   Benson, R. C.; Meyer, R. A.; Zaruba; M.E, McKhann G.M. *J.   Histochem. Cytochem.* **1979**, *27*, p. 44.
(7)   Aubin, J. E. *J. Histochem. Cytochem.* **1979**. *27*, p.36.
(8)   Shapiro, H.M. *Practical Flow Cytometry*, 2nd Ed., Alan R. Liss, Wiley; New York, N.Y, **1988.**
(9)   Loken,   M.R.;   Parks,   D.R.;   Hertzenberg,   L.A.   *J. Histochem. Cytochem.* **1977**, *25*, p. 899.
(10) Glazer, A.N.; Stryer, L. *Biophys J.* **1983**, *43*, p. 383.
(11) Leif, R.C.;   Clay, S.P.;   Gratzner, H.G.;   Haines, H.G.; Rao K.V.; Vallarino, L.M. In *The Automation of Uterine Cancer Cytology*, Wied, G.L.; Bhar, G.F.; Bartels, P.H. Eds.; Tutorials of Cytology; Chicago, IL, **1976**; p. 313.
(12) Crosby, G.A. *Molecular Crystals*, **1966**, *1*, p. 37.

(13) Bhaumik, M.L. *J. Chem. Phys.* **1964** *40*, p. 3711.
(14) Soini, E.; Lovgren T.   In *CRC Critical Reviews in Analytical Chemistry*, **1987**, *18*, Issue 2, p. 105.
(15) Vallarino, L.M.; Watson, B.D.; Hindman, D.H.K.; Jagodic V.; Leif R.C. In *The Automation of Cancer Cytology and Cell Image Analysis*, Pressman, H.J.; Wied, G.L. Eds.; Tokyo, **1979**; p. 53.
(16) Leif, R.C. In *Automated Cell Identification and Sorting*, G. L. Wied and G. F. Bahr, Eds, Academic Press, New York, **1970,** p. 131.
(17) Mathies, R. A; Stryer, L. In *Applications in Biomedical Sciences*, Alan R. Liss, Inc. **1986**, p. 129.
(18) Soini, E.; Hemmila, I.   U.S. Patent 4,374,120, **1983**.
(19) Mikola, H.; Mukkala, V.-M.; Hemmila, I., WO Patent 03698, **1984.**
(20) Hemmila, I.; Dakubu, S.; Mukkala, V.M.; Siitari, H.; Lovgren, T. *Anal. Biochem*, **1984**, *137*, p. 335.
(21) Evangelista, R.A.; Pollak, A.; Allore, B.   *Clin. Biochem*, **1988**, *21*, p. 173.
(22) Khosravi, M.J.; Diamandis, E.P. *Clin. Chem.* **1989**, *35*, p. 181.
(23) Soni, E.J; Pelliniemi, L. J.; Hemmila, I. A.; Mukkalva, V-M, Kankare, J. J.; Frojdman, K. *J. Histochem Cytochem*, **1988**, *36*, p. 1449.
(24) Beverloo, H. B.; van Schadewijk, A.; van Gelderen-Boele, S.; Tanke, H. J. *Cytometry*, **1990**, *11*, p. 784.
(25) Filipescu, N.; Sager, W.F.; Serafin, F.A. *J. Phys. Chem.*, **1964**, *68*, p. 3324.
(26) Dutt, N.K.; Bandyopadhyay, P. *J. Inorg. Nucl. Chem.* **1964**, *26*, p.729.
(27) Gent, N.J. Determination of Stability Constants of Lanthanide Complexes by Fluorescence Spectroscopy, M.S. Thesis, Virginia Commonwealth University, 1983.
(28) Melby, L.R.; Rose, N.J.; Abramson, E.; Caris, J.C. *J. Am. Chem. Soc.*, **1964**, p. 5117.
(29) McGuire, A.A.P.; Vallarino, L.M. *30th South Eastern Regional Meeting*, American Chemical Society; Savannah, GA, Nov. 1978; Abstr. 152
(30) Melson, G.A. in *Coordination Chemistry of Macrocyclic Compounds*, Melson, G.A. Ed.; Plenum; New York, **1979**; p. 17.
(31) Backer-Dirks, J.D.J.; Gray, C.H.; Hart, F.A.; Hursthouse, M.B.; Schoop, B.C. *J. Chem. Soc. Chem. Commun.* **1979**, p.774.
(32) Smailes, D.L.; Vallarino, L.M. *Inorg. Chem.* **1986**, *25*, p. 1729.
(33) Sabbatini, N.; De Cola, L.; Vallarino, L.M.; Blasse, G.J. *J. Phys. Chem*, **1987**, *91*, p.4681.
(34) Smailes, D.L. *Hexa-aza Macrocyclic Complexes of the Lanthanides: Synthesis and Properties*, M.S. Thesis. Virginia Commonwealth University, **1986.**
(35) De Cola, L.; Smailes, D.L.; Vallarino, L.M. *X Convegno Nazionale di Fotochimica*, Ravenna, Italy, **1985**; p. 4.
(36) Gootee, W.A; Pham, K.T.; Vallarino, L.M. *41st South Eastern Regional Meeting*, American Chemical Society, Winston-Salem, NC., Nov. **1989**; Abstr. 313.

(37) Gootee, W.A. *Study of Metal Ions in Polymers: Model Compounds for the Inclusion of Lanthanides in Polyimides and Bifunctional Lanthanide(III) Macrocyclic Complexes*, Ph.D. Dissertation, Virginia Commonwealth University, **1989**.
(38) Gribi, C.; Smailes, D.L.; Twiford, A.; Vallarino, L.M. *41st South Eastern Regional Meeting*, American Chemical Society, Winston-Salem, NC., Nov. **1989**; Abstr. 312.

RECEIVED March 15, 1991

Chapter 4

Automated Cell Separation Techniques Based on Optical Trapping

T. N. Buican

Life Sciences Division, Los Alamos National Laboratory,
Los Alamos, NM 87545

Optical trapping can be used to levitate and manipulate a wide variety of microscopic particles, including living cells and chromosomes in aqueous suspension. Both two-dimensional (2-D) and three-dimensional (3-D) optical traps can be easily produced and can be used, respectively, for sorting and manipulating microscopic particles. We describe two cell separation techniques developed in our laboratory: (1) laser sorting, based on the use of 2-D traps; and (2) microrobotic manipulation, which uses a 3-D optical trap, video microscopy and machine vision in order to separate single cells and chromosomes. Both techniques can be integrated into complex instruments for the analysis, separation, manipulation and processing of individual cells and cell organelles.

Optical trapping, a purely optical technique for the manipulation of microscopic particles, was invented in the late '60s by Arthur Ashkin of AT&T Bell Labs (1). The technique relies on the pressure created by one or more laser beams that are scattered by a microscopic object in order to trap, levitate, and move that object. As opposed to other trapping techniques, optical traps are intrinsically stable and very localized in their effects. As such, they can be incorporated into relatively simple devices that allow single cells, chromosomes, and other cell organelles to be accurately positioned and transported. Furthermore, optical trapping only requires low-intensity laser beams and can be operated at wavelengths at which absorption by the trapped particle is minimized. Thus, for most cells that have been optically trapped, the trapping laser beams seem to have negligible biological effects (2-4).

Cell separation based on optical trapping belongs to the class of *active, single-cell* sorting techniques, together with flow sorting and mechanical micromanipulation. As such, automated optical trapping instruments analyze single cells and, after converting the measurement results into a sorting decision, physically remove selected single cells from the original sample. Although passive cell separation based on optical trapping has also been proposed (5), the complex dependence of separation properties on the physical properties of the cell and the existence of beam-mediated interactions between cells (6) make this technique unattractive. By contrast, active separation based on optical measurements of single-cell prop-

0097–6156/91/0464–0059$06.00/0

erties and automated beam control can be very precise both in terms of its analysis of single cells and in the physical separation of these cells. Furthermore, automation applies in a natural way to optical trapping as the trapping optics can easily be used for light collection and imaging, and the direction and intensity of the laser beams used for trapping can easily be brought under automated control.

The range of cell properties on which automated cell separation can be based in an optical trapping system includes labelling fluorescence intensity, light scattering intensity distribution, and cell morphology. As the speed at which cells travel through the analysis volume can be accurately controlled and is much lower than the typical speeds encountered in flow sorting, high-resolution measurements of cell properties can be made. Furthermore, since cell position can be controlled with very high accuracy, other, nonoptical measurements can be performed.

The purely optical nature of cell manipulation by optical trapping means that cell analysis and manipulation can be performed inside *completely enclosed* sample systems. Thus, separation based on optical trapping has a clear advantage over mechanical micromanipulation, which can only be performed in open containers. The small diameter of the trapping beams and the highly localized character of the optical trap allow manipulation to be performed inside commensurately small compartments. Consequently, a single optical manipulation chamber can contain a large number of compartments and interconnecting channels. The former can be used not only for collecting cells that have been separated, but also for performing further analysis and processing of those cells. Thus, complex experimental protocols can be carried out at the level of the single cell inside an optical manipulator, without the need for recovering the cells after each step and transporting them between separate instruments.

We begin this article with a brief presentation of the principles of optical trapping. Following this, we discuss the applications of two-dimensional (2-D) and three-dimensional (3-D) optical traps in cell separation. For each type of trap we discuss instrument optics, manipulation chambers, control electronics, and the procedures for automated control. We conclude with a brief discussion of future developments in the field of cell separation and processing by optical trapping.

A Qualitative Discussion of Optical Trapping

Whenever light is scattered by an object, the change in the momentum of the scattered photons leads to a momentum of equal magnitude but opposite direction being transferred to the scattering object. If one considers a system consisting of a scattering object and the incident and scattered beams (Figure 1), and if P_0 and P_1 are, respectively, the momentum fluxes (momentum transported by the beams per unit time) of the incident and scattered beams, then, because of conservation of total momentum, a force, $F = P_1 - P_0$, is produced that acts on the scattering object. This is the *radiation pressure force*, and we will refer in what follows to its components along the direction of the incident beam (axial force) and perpendicular to that direction (radial force).

The momentum carried by a light beam in a given direction can be computed from the far-field angular distribution of the beam intensity, ρ, and is described by

$$P = \frac{W}{c} \int_{4\pi} \rho \cos\alpha \, d\Omega$$

where W is the total beam power, c is the speed of light, α is the angle relative to the given direction, and $d\Omega$ is the solid angle element (7). For an axially symmetric beam, the factor, $\cos\alpha$, shows that, at constant total beam power, beam mo-

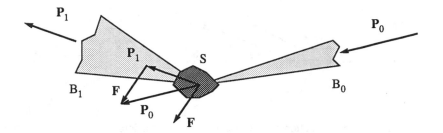

Figure 1.   Conservation of momentum in a light scattering system: B_0, B_1—incident and scattered beams; S—scattering object; P_0, P_1—momentum fluxes of the two beams;   F—light pressure force due to scattering of the incident beam by S.

mentum decreases as the divergence of the beam increases. It follows immediately that beam momentum is decreased by (i) absorption (which reduces beam power); and (ii) diffuse scattering (which spreads out the intensity distribution). Absorption produces a purely axial force in the direction of the incident beam. Diffuse scattering, if symmetric relative to the direction of the incident beam, also contributes a force in the direction of the incident beam.

It can be shown that the momentum flux of a beam with axial symmetry is parallel to the beam axis and has a magnitude given, for small θ, by $P = W (1 - \theta^2) /c$, where θ is an angular parameter describing beam divergence. This relationship shows that the beam has maximum momentum when it is maximally collimated ($\theta = 0$), and its momentum decreases as its divergence increases. The magnitude of the force acting on the scattering particle is thus given by $F = - W \Delta(\theta^2) / c$. If the scattering object behaves like a lens or mirror, the beam angle, θ, can either increase or decrease, depending on the geometry of the object and incident beam. Consequently, we may expect both the magnitude and direction of the light pressure force acting on an optical element to depend on the optical properties of the object, as well as its position relative to the incident beam.

The Lens Model. Under most experimental conditions, live cells have a refractive index close to that of the suspension medium (*8*) and, therefore, at least in a first approximation, partial reflection does not play an important role in biological optical trapping. An adequate qualitative description of the optical trapping of biological particles can thus be obtained by treating the particles as lenses that refract the incident beam.

Radial Trapping. It is easy to see that, whenever an object that behaves like a convergent lens (internal index of refraction higher than that of the surrounding medium) is moved away from the axis of a light beam incident on it, the light beam is deflected away from the initial direction (Figure 2a). Therefore, the radial component of the light pressure force acting on the object points in a direction opposite to particle displacement (toward the beam axis), and a stable, 2-D trapping effect is created. One can similarly show that the radial force acting on a divergent lens (object with a lower refractive index than that of the medium) points in the same direction as the displacement (Figure 2b) and thus there is no radial trapping effect. Radial trapping applies to both spherical and nonspherical

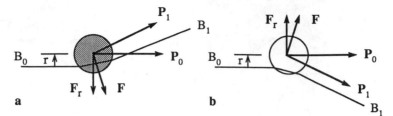

Figure 2. The orientation of the radial force: (a) particle with higher refractive index than that of the surrounding medium; and (b) particle with lower refractive index. B_0, B_1—incident and scattered beams; F—light pressure force; F_r—radial component of F; P_0, P_1—momentum fluxes for the incident and scattered beams; r—radial particle displacement. Only the axes of the beams are shown.

particles. In the case of the latter, torques may develop which give the particles preferential orientations relative to the incident beam.

Axial Trapping. In the lens model of optical trapping, the particle in the laser beam is regarded as a lens of focal length, f (Figure 3). In a purely geometrical optics approximation, one can easily see that there are always two points along the beam axis for which the angle of the output ("scattered") beam equals that of the

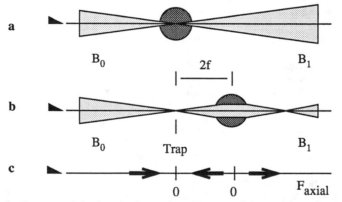

Figure 3. Lens model of optical trapping. Two particle positions for which the axial force vanishes: (a) particle at beam focus; (b) particle at a distance 2f from the beam focus; (c) diagram showing the direction of the axial force (F_{axial}) for various positions of the particle. The position of the stable axial trap relative to the trapping beam is indicated. The focal length of the particle is denoted by f. The incident and scattered beams, B_0 and B_1, travel from left to right.

incident beam. Thus, the momentum fluxes of the two beams have the same magnitude and, for distances of 0 and 2f between the lens and the beam focus, the axial force vanishes (Figure 3c). It is also easy to see that, for any distance between 0 and 2f, the beam angle decreases after passing through the object and, therefore, the beam momentum increases and the light pressure force acting on the lens is directed opposite to the beam. For all points outside this interval, the beam momentum decreases and the lens is pushed in the direction of the beam. As a consequence of this dependence of the axial force on the distance between lens (object) and beam focus, a stable axial trap always exists in a purely geometrical optics model and is situated at the beam focus. However, if one considers a Gaussian beam and the corresponding imaging relationships (9), one can show that an axial trap exists only if $\theta^2 > \lambda/\pi f$, where λ is the wavelength and θ is the asymptotic angle of the Gaussian beam (9). Thus, for weakly convergent beams (small θ), the inequality above is not satisfied. Consequently, the axial force always points in the direction of the beam and there is no axial trapping. One can thus define two trapping regimes, depending on the degree of convergence of the beam: (i) 2-D trapping, which applies to weakly convergent beams and in which there is a trapping effect only in a plane perpendicular to the axis of the trapping beam; and (ii) 3-D trapping, in which a strongly convergent laser beam creates a full, 3-D optical trap.

2-D Optical Trapping and Cell Separation

We showed in the previous section that weakly collimated laser beams produce 2-D optical traps and that trapped particles are propelled along the beam axis. Such traps may be used to transport microscopic particles over macroscopic distances and can thus be the basis for both remote delivery and sorting systems.

As the particles in a 2-D trap move along the beam axis, hydrodynamic forces which depend on particle shape and size will affect both the velocity at which the particles travel and the stability of the trapping effect. Thus, during the 2-D trapping of blood cells, we observed that lymphocytes, which are roughly spherical, are stably trapped and can be transported over relatively large distances (6). By contrast, erythrocytes are only briefly held in the beam and are readily pushed out by orientation-dependent hydrodynamic forces. This phenomenon could be used for the passive separation of spherical from nonspherical cells.

The velocity at which stably trapped cells travel along weakly focused beams depends on both the optical and the hydrodynamic properties of the cells. The use of differences in travel velocity for passive cell separation has been proposed by Ashkin (5). However, the travel velocity of a given particle is further affected by the scattering of the beam by closely following, trapped particles. We have reported (6) that the interaction between simultaneously trapped particles may lead to the formation either of particle clumps or of stable systems of particles which move at the same velocity while maintaining a constant separation. In either case, passive separation is no longer possible, as travel velocity now depends on the properties of more than one cell. The simultaneous presence of several particles in the trapping beam can be avoided by lowering the rate of particle injection and thus limiting the throughput of any passive sorting system of this kind.

The use of weakly focused laser beams to transport, rather than sort, microscopic particles has the important advantage that proper alignment of the transportation beam with the collection volume is sufficient to ensure precise delivery of the trapped particles. Thus, a transportation system based on a weakly collimated beam requires minimal adjustment. Furthermore, such a system can accurately deliver microscopic particles through apertures with diameters comparable with the diameter of the trapping beams, and can thus move selected particles between separate compartments in a complex, integrated instrument.

Laser Sorting. The macroscopic transportation capabilities of weakly focused laser beams can be used for the automated sorting of microscopic particles. A particle travelling along a weakly focused beam can be transferred to another, intersecting beam, provided the intensity of the latter is above a threshold value that depends on the intensity of the first beam and the relative positions of the beam

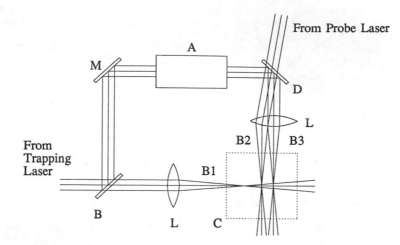

Figure 4. Simplified diagram of the laser sorter optics: A—acoustooptic modulator; B—beam splitter; B1—propulsion beam; B2—probe beam; B3—deflection beam; C—manipulation chamber; D—dichroic mirror; L—focusing lenses; M—mirror. The light collection optics are not shown.

waists. Intersecting beams can thus be used to direct trapped particles to one of several collection volumes. An automated sorting system based on light scattering measurements is described in (6) and a diagram of the instrument optics is shown in Figure 4. Naturally, other optical, as well as nonoptical, parameters can be used to control sorting. A manipulation chamber for laser sorting (6) is shown in Figure 5. Channels machined into the top of the chamber are enclosed by a window that allows the particles in the chamber to be visualized. Light scatter measurements are also made through this window. The laser beams enter the channels through side windows, and the sample is injected into the propulsion beam through a sample port. There are two ports just ahead of each side window that are used to create localized fluid flows. Individual cells are trapped in the propulsion beam and start travelling along its axis. As each cell intersects the probe beam (of wavelength different from that of the trapping beams and of lower inten-

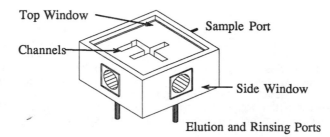

Figure 5. View of the laser sorter manipulation chamber (see Figure 4 for the geometry of the beams inside the chamber). The side windows allow the three laser beams to enter and exit the chamber. The elution and rinsing ports are used to create local fluid flows near the side windows.

sity), the scattered light intensity is measured and a sorting decision is made. When a cell reaches the deflection beam (typically about 100 μm downstream from the probe beam), the intensity of this beam is modified by an acoustooptic modulator so as to direct the particle to the appropriate elution volume. Localized fluid flows just ahead of the beam output windows are used to elute the cells.

The beam focusing optics on the laser sorter described in (6) had a focal length of 50 mm. At a power of approximately 300 mW per beam, typical particle velocities reached several hundred μm/s. For a separation between probe and deflection beams of 100 μm, this resulted in a maximum sorting frequency of a few cells per second. Although this rate is too low to make the instrument a practical sorter, its simplicity and the relatively large distances travelled by the cells (over 6 mm) may allow a similar system to be used as a microscopic gating device that prevents unwanted cells from drifting through narrow apertures connecting the compartments of an integrated analysis and preparation instrument, while letting through selected cells.

The elements of the sorting control system are shown in Figure 6. The light

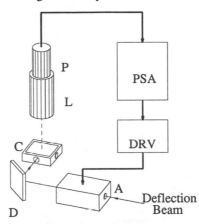

Figure 6. Automated control of the laser sorter: A—acoustooptic modulator; C—manipulation chamber; D—dichroic mirror; DRV—modulator driver; PSA—pulse shape analyzer; L—lens; P—photomultiplier. The dashed line represents the light collection path, while the wide, solid line represents the signal and control path. The deflection beam focusing lens is not shown.

scattered by each cell as it passes through the probe beam is collected by a lens and falls on a photomultiplier. The pulse produced by the detector is analyzed by a pulse shape analyzer and a sorting decision is made on the basis of the properties of this pulse.

The parameter of choice for controlling the laser sorter is pulse amplitude. Pulse area, which is a commonly used sort control parameter in flow sorters, is not suitable for controlling laser sorting because pulse widths are not constant. This is due to the fact that particle velocity in the propulsion beam depends on many particle properties and is also affected by the presence of other particles in the beam (6).

Secondary parameters, such as pulse width and time interval between pulses, are also derived by the pulse shape analyzer. These parameters, the values of which can indicate a separation between successive particles too small to allow individual analysis and/or separation, are used to disable deflection and thus avoid compromising sort purity.

Microrobotic Manipulation

A different approach to cell separation is made possible by the use of a 3-D optical trap. While the laser sorter is operationally equivalent to a flow sorter (in the sense that it separates cells into a small number of collection volumes and that measurements are performed on particles moving along a well-defined axis), the 3-D optical trap can be used for microscopic robotic manipulation of single cells and cell organelles. This technique is well suited for the automated separation of rare cells and can also be easily integrated into complex instruments.

Optical Manipulator Designs. Two basic optical manipulator designs, both originating in Ashkin's work (*1, 10*), have been proposed and implemented. The two designs differ in the number of laser beams used to produce a 3-D optical trap (Figure 7). The single-beam optical manipulator (Figure 7a) relies upon the use

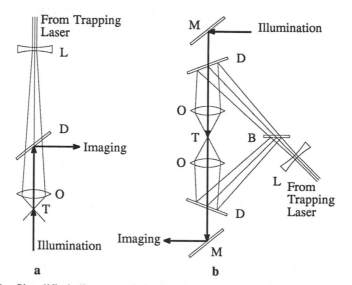

a   b

Figure 7. Simplified diagram of single-beam (a) and double-beam (b) optical trapping and imaging systems. B—beam splitter; D—dichroic mirror; L—divergent lens; M—mirror; O—microscope objective; T—optical trap. The arrows represent the imaging optical paths.

of a high numerical aperture focusing lens and has been implemented on commercial microscopes (*10*). The double-beam system (Figure 7b) uses lower numerical aperture lenses, and a stable 3-D trap is produced by two coaxial, counterpropagating beams (*11*). In both cases, the lenses used for focusing the trapping beam(s) are also used for imaging, and dichroic mirrors are used to separate the trapping and imaging wavelengths.

The single beam optical trap illustrated in Figure 7a can be implemented on any microscope with only minor modifications to the microscope optics. The illumination source is the microscope condenser, while imaging is done through the existing microscope optics. The role of the additional divergent lens is to expand the trapping laser beam in order to fill the input aperture of the microscope objective and thus maximize the convergence of the trapping beam. The position of the trap within the field of view of the microscope can be modified by moving the divergent lens (*3*). Longer-range motion of the trap can be achieved by moving the manipulation chamber relative to the trapping beam. A variant of Ashkin's single-beam design was proposed in (*12*), where an optical trap was added to a confocal microscope by means of a second microscope objective. The second objective, which is used exclusively for trapping, is mounted on an xyz stage and can thus control the position of the optical trap.

The trapping beams in the double-beam system are derived from a single laser beam by means of a beam splitter. The use of a divergent lens for beam expansion is optional, as a high degree of convergence of the trapping beams is no longer critical in this system. The optics for this type of manipulator are more complex than those of the Ashkin design and attention must be paid to the proper alignment of the trapping beams. However, beam alignment can be monitored by a video camera that images the beams after a complete round trip through the system and can thus be automated (*11*).

Because of their different optical characteristics, the two optical manipulator designs have different applications. The single-beam manipulator, which uses a high numerical aperture lens, is capable of high-resolution imaging and can produce small-diameter, high-intensity traps. This manipulator is therefore most suitable for the analysis and manipulation of small particles such as cell organelles. At the same time, the use of a high numerical aperture lens leads to a small field of view and a very small working distance. The small field of view restricts the use of such a manipulator in an automated separation system, where the performance of automated navigation algorithms depends to a large extent on the size of the field of view. Furthermore, the very small working distance makes difficult the use of complex manipulation chambers.

By contrast, the dual beam system provides lower-resolution imaging but has a wider field of view and larger working distance. This system is eminently suited for automated cell and chromosome separation. In applications such as chromosome separation, the lower resolution of the images provided by the double beam manipulator can be compensated for through the use of specific fluorescent probes. The large working distance which, for modern ultralong working distance objectives, can be around 1 cm, allows the optical manipulator to be easily integrated into complex systems.

Optical Manipulation Chamber Designs. Optical manipulation chambers may incorporate features that facilitate cell separation. For instance, multiple compartments inside the chamber allow selected cells to be physically separated from the original cell sample. It is also desirable for the manipulation chamber to have ports through which the sample can be introduced and the separated cells can be eluted. If the chamber has multiple compartments, then it is preferable for each compartment to be connected to separate external ports. Finally, if the cell sample is highly concentrated (as is the case, for example, with blood samples), it is desirable for the manipulation chamber to be shallow. Low chamber depth ensures that, after sedimentation of the cells, the surface concentration is sufficiently low to permit the trapping and removal of single cells.

Manual cell separation through optical manipulation was demonstrated by Ashkin (*3*) in a rather crude chamber that contained a hollow, transparent fiber. We described a multicompartment chamber design (*11*) that is well suited for cell separation, while also containing features that make it attractive for complex work with live cells. This chamber (Figure 8) consists of three layers that are cemented together. The outer layers are good quality glass of 170-μm and 1-mm thickness respectively. The central layer, which contains the pattern of compartments and channels, can be fabricated either from stainless steel shim stock or from a photosensitive ceramic (Fotoceram, Corning). Small compartments and channels can be accurately cut through the central layer by laser machining (stainless steel) or

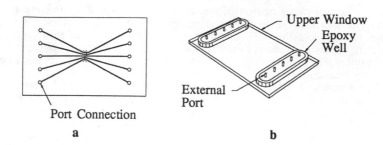

Figure 8. Multicompartment manipulation chamber: (a) Top view of the central layer, showing compartments (central region) and interconnecting channels; and (b) perspective view of the upper window, showing the external ports.

photoetching (Fotoceram). A view of the central section is shown in Figure 8a, where one can see five compartments (300 μm by 300 μm) and channels connecting the compartments with each other and with the external ports. The external ports are connected to internal channels through holes in the 1-mm-thick glass window, to which they are cemented (Figure 8b). The small thickness and large surface area of this manipulation chamber lead to efficient heat exchange and temperature control. This characteristic is particularly important when separating and processing live cells, as well as for performing biochemical assays inside the chamber. Furthermore, the channels connecting each compartment to external ports allow reagents and culture media to be circulated through the compartments, and complex, long-term protocols can thus be carried out inside the chamber.

Automated Control Based on Machine Vision. The ability of optical manipulators to image and manipulate single cells makes this type of instrument amenable to machine-vision-based automation. The control problem in such an instrument consists of the following: (i) finding the cells of interest; and (ii) separating those cells without collecting unwanted particles. In order to simplify image acquisition, we assume that the particles injected into the chamber have been allowed to settle onto the bottom window. Obviously, a shallow chamber has the advantage that particle settling occurs quickly and the delay between sample injection and the beginning of the search procedure is short. The search procedure consists of scanning through the sample and analyzing each particle in the instrument's field of view. Once a cell of interest has been identified, it is trapped and raised to a certain height above the bottom window (Figure 9), following which it is transported to an area where it is either recovered or further analyzed and processed.

Given the initial position of the trapped cell and the coordinates of the target position, separation must proceed along a path that does not collide either with the chamber walls or with unwanted particles. For a simple chamber consisting of a single, convex compartment, the first requirement is easily satisfied. The simplest path satisfying the second requirement depends on the depth of the chamber. As optical traps are generally better localized in a plane perpendicular to the trapping beams than along the beam axis, a shallow chamber may require the trap to be moved around unwanted particles in order to avoid trapping them. This is illustrated in Figure 9, where we distinguish between the *separation path* followed by the trap, and the *clear path*, which is the projection of the former onto the bottom

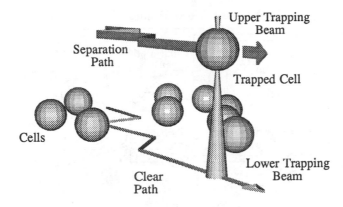

Figure 9. Geometry of cell separation by optical manipulation. The trapped cell is carried along a separation path that avoids other cells on the bottom of the manipulation chamber. The cells in the chamber are imaged along the common axis of the trapping beams. The clear path, which is the projection of the separation path onto the bottom of the chamber, is constructed so as to avoid the trapping of unwanted cells.

of the chamber. Finding a clear path from a knowledge of the positions and sizes of the settled cells is the classical collision avoidance problem of robotics. In the case of an optical manipulator, a clear path must be computed gradually, on the basis of a succession of overlapping video frames (Figure 10). This will be discussed in more detail in the next section. Automated navigation is not required for chambers that are deeper than the axial dimension of the trap, and the separation path is, consequently, a straight line at a height above the bottom window that is sufficient to prevent the trapping of unwanted cells.

The computation of the separation path is slightly more complicated for a multicompartment chamber, where the trap must avoid not only other cells, but also the edges of compartments and channels. In this case, a rough path that avoids collision with chamber edges may have to be computed first, following which the rough path is used to define the local direction of motion followed by the clear path (Figure 10). The rough path can be easily computed from a stored map of the manipulation chamber (*11*).

Single Video Frame Navigation. The automated navigator that controls cell separation by optical manipulation must find a clear path within the boundary of the current field of view, such that (i) it avoids collision with particles on the bottom of the chamber; and (ii) it leads, as closely as possible, in the direction defined by the rough path. Once such a path has been found for the current field of view, the instrument moves the optical trap and the trapped particle along the clear path to a point close to the edge of the video frame (Figure 10). Following this, a new video frame is acquired and the process is repeated until the target position is reached.

A wide variety of algorithms for finding collision-free, clear paths have been constructed (*13*). We have developed a navigator algorithm that is derived from the circle tangency techniques described in (*14*). This type of navigator algorithm, which approximates objects by circles of appropriate radius, is computationally efficient and is well suited for the usual cell shapes. The circle tangency algorithm

Successive Video Frames

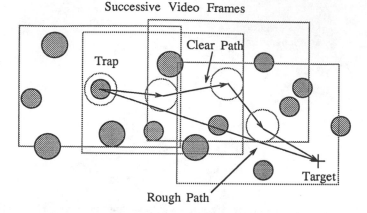

Figure 10. Automated navigation for robotic particle separation: Computation of a collision-free, clear path through a succession of overlapping video frames. The point labelled "Target" is, depending on the location of the trap along the rough path, either the next vertex of the latter, or the final target position.

constructs a clear path that consists of arcs along circular object boundaries and line segments that are tangent to the boundary of the last encountered object and point in the direction of the current target point. If the target point is not in the current video frame, all clear paths leading to the frame boundary are computed, together with weights that increase with path length and deviation from the direction leading directly to the target. The path with the lowest weight is chosen, the trap is moved to the point on the path within a trap radius of the boundary, a new video frame is grabbed, and the process is repeated until the target position is reached.

The navigator algorithm stores the position and size of all previously detected objects and uses this information when computing a path. This ensures that no path is chosen that leads to a previously detected object that lies beyond the boundary of the current frame. The algorithm can also move back along a path if it finds itself inside a cul-de-sac. The algorithm monitors the continued presence of a particle inside the trap and alerts the operator if the particle is lost.

The example in Figure 11 illustrates the operation of the automated navigator and shows the ability of the algorithm to use previously acquired information. The diagrams are printouts of the image display area of the user interface screen, showing polystyrene microspheres with a diameter of 5.8 μm. The x, y coordinates (in μm) of the optical trap and of the field of view are shown, respectively, near the bottom and left sides of the images. The rough path, shown in Figure 11a, intersects particle 2. The navigator algorithm computes clear paths within the current video frame and chooses the path shown in Figure 11b as having the smallest angular deviation from the rough path (the clear path extends from the trap to the stopping point, S). The position information for the other particles in the frame is stored. As the trap is moved to the stopping point of Figure 11b, a new frame is grabbed (Figure 11c). A new computation of the clear paths is performed, and the navigator reverses direction by using the previously stored information about the position of particle 1 (Figure 11d). The trap is successfully moved around particle 1 (Figure 11e), and the target is reached in Figure 11f.

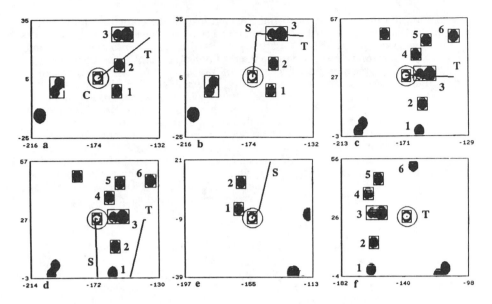

Figure 11. Automated navigation of the optical trap. The binary images show polystyrene microspheres with a diameter of 5.8 μm. One particle is held in the optical trap (central circle). The trapped particle must be moved without collision from the current position, C, to the target point, T. C—current trap position; S—stopping point; T—target position; 1-6—obstacles.

Automated Control System. The main elements of the control system for microrobotic cell separation are shown in Figure 12. The imaging system, which is no more than a video microscope, provides a video signal which is digitized by a frame grabber and made available to a processor. The processor performs a series

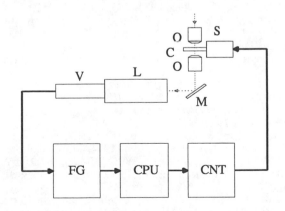

Figure 12. Automated control of the optical manipulator: C—manipulation chamber; CNT—servo controller; CPU—central processing unit; FG—frame grabber; L—imaging lens; M—mirror; O—microscope objective; V—video camera; S—xyz stage with servo actuators. The dotted line represents the imaging optical path, while the solid line represents the data and control path.

of operations on the digitized video frame which lead to the computation of a local clear path. Once the clear path for the current frame has been computed, the processor downloads the coordinates of the path vertices to a servo controller, and the servo actuators on an xyz stage move the manipulation chamber accordingly.

Given the computation-intensive nature of the image processing and machine vision algorithms, it is obvious that the speed at which cells can be separated depends, to a large extent, on the computing power available for controlling the optical manipulator. As shown in Figure 12, the robotic manipulator described in (11) uses a single processor both for running the image analysis and machine vision algorithms, and for controlling the relative position of the optical trap. Furthermore, the same processor operates the rest of the hardware, as well as a high-resolution, interactive user interface. Consequently, separation of trapped cells does not proceed at the maximum possible speed, which is only limited by viscous drag and beam intensity. The data processing bottleneck can be overcome by using real-time, multiprocessor systems, which are now available at reasonable cost.

Conclusions

The manipulation techniques made possible by optical trapping are well suited for large-scale automation and integration. As such, they have the potential to provide the technological basis for complex, powerful, and efficient "microscopic laboratories" for cell biology, cytogenetics, and molecular biology. Such instruments would perform complex analytical and preparative procedures on single cells and chromosomes, while requiring a minimum of human intervention. This approach may open up new avenues in experimental biology and biotechnology.

Acknowledgments

The automated navigator software was written by Bryan D. Upham. This work was supported in part by LANL ISRD Award X86C, NASA contract T-1196P, and by the National Flow Cytometry Resource (grant RR01315).

Literature Cited

1. Ashkin, A. *Phys. Rev. Lett.* **1970,** *24,* pp. 156-159.
2. Ashkin, A.; Dziedzic, J. M. *Science* **1987,** *235,* pp. 1517-1520.
3. Ashkin, A.; Dziedzic, J. M.; Yamane, T. *Nature* **1987,** *330,* pp. 769-771.
4. Block, S. M.; Blair, D. F.; Berg, H. C. *Nature,* **1989,** *338,* pp. 514-518.
5. Ashkin, A. *U. S. Patent No. 3,710,279,* 1973.
6. Buican, T. N.; Smyth, M. J.; Crissman, H. A.; Salzman, G. C.; Stewart, C. C; Martin, J. C. *Appl. Opt.* **1987,** *26,* pp. 5311-5316.
7. Bohren, C. F.; Huffman, D. R. *Absorption and Scattering of Light by Small Particles;* John Wiley & Sons: New York, NY, 1983; p. 72.
8. Salzman, G. C. In *Cell Analysis;* Catsimpoolas, N., Ed.; Plenum Publishing Co.: New York, New York, 1982; Vol. 1; pp. 111-143.
9. Kogelnik, H.; Li, T. *Appl. Opt.* **1966,** *5,* pp. 1550-1567.
10. Ashkin, A.; Dziedzic, J. M. *Appl. Phys. Lett.* **1974,** *24,* pp. 586-588.
11. Buican, T. N.; Neagley, D. L.; Morrison, W. C.; Upham, B. D. In *New Techniques in Cytometry;* Salzman, G. C., Ed.; Proceedings of SPIE; SPIE: Bellingham, Washington, 1989, Vol. 1063; pp. 190-197.
12. Visscher, K.; Brakenhoff, G. J. *Cytometry* **1990,** Supplement *4,* p. 17.
13. Horn, B. K. P. *Robot Vision;* The MIT Electrical Engineering and Computer Science Series; MIT Press, Cambridge, MA, 1986.
14. Moravec, H.P. *Obstacle Avoidance and Navigation in the Real World by a Seeing Robot Rover;* Ph.D. dissertation; Stanford University, Palo Alto, CA, 1980.

RECEIVED April 15, 1991

Chapter 5

Separation Techniques Used To Prepare Highly Purified Chromosome Populations

Sedimentation, Centrifugation, and Flow Sorting

K. L. Albright, L. S. Cram, and J. C. Martin

Life Sciences Division, Los Alamos National Laboratory,
Los Alamos, NM 87545

Flow Cytometry analyzes heterogeneous chromosome suspensions and electronically separates desired chromosome types as pure fractions. Adjunct techniques have been used to improve the efficiency of flow sorting. Prefractionation of the initial chromosome suspension has been carried out on sucrose gradients. Concentration of dilute suspensions of mitotic chromosomes has been accomplished using sucrose and Metrizamide interfaces and agarose plugs.

Chromosome sorting using flow cytometry is a primary technique for the construction of chromosome specific DNA libraries (1). The purpose of this chapter is to describe techniques used at Los Alamos National Laboratory for high speed chromosome sorting, and new approaches for fractionating chromosome preparations prior to high speed flow sorting.

Whole genomic DNA can be used to make libraries (a set of cloned DNA fragments) by any of the usual molecular biology techniques. A major drawback of genomic libraries is the vast number of irrelevant clones that have to be screened to analyze a specific chromosome or a subregion of a chromosome. This drawback can be circumvented by constructing libraries using DNA of one chromosomal type. A major Department of Energy program was established in 1983 at the Los Alamos and Lawrence Livermore National Laboratories to purify human chromosome specific DNA by flow sorting and to construct libraries for each of the 24 chromosome types. Two complete sets of small insert libraries have been constructed (2) and are available through the American Type Culture Collection (3). Larger insert libraries are being constructed in cosmid and bacteriophage vectors and will soon become available. Even larger insert clones, e.g. yeast artificial chromosome libraries (300 kb), are currently being investigated.

The recovery of sorted chromosomes and the molecular weight of the extracted DNA needs to be maximized. The molecular weight of sorted chromosomal DNA is currently >1 Mbase (10^6 nucleotides). Improvements in both of these areas of are under development at Los Alamos National Laboratory and other labs.

0097–6156/91/0464–0073$06.00/0

Chromosome Preparation

Many different methods have been developed to isolate chromosomes. Our interest lies with bulk preparation techniques which require 10^6 to 10^7 mitotic cells and will result in 10^8 to 10^9 free mitotic chromosomes.

The methods we have used for isolating single chromosomes include: the Aten technique for single parameter flow karyotyping (4), the Wray-Stubblefield hexalene glycol technique (5), which is applicable to single and dual parameter flow karyotyping and has the potential for banding of sorted chromosomes, the magnesium sulfate method which gives good resolution and low debris levels for two parameter karyotypes (6), and the polyamine method of Sillar and Young which yields the highest molecular weight DNA of all these techniques and generally gives good resolution and low-debris two-parameter flow karyotypes (7).

Cell Culture. Chromosomes have been isolated from many cell types, such as human diploid fibroblasts, human lymphoblasts, and hybrid cell lines of mouse or hamster origin which contain one or only a few human chromosome types. Fibroblasts are grown as adherent monolayers and lymphoblasts are grown in suspension culture. Hybrids are usually grown as a monolayer, but can sometimes be grown in suspension culture. A large number of different cell lines have been screened for suitability for chromosome sorting (2). Each cell line has its own unique growth conditions and preparation requirements.

Chromosome Isolation. The polyamine method is preferred for preserving the molecular weight of chromosomal DNA and has become the standard method for chromosome preparation for the National Laboratory Gene Library project. For adherent cell cultures, mitotic cells can be selectively separated from the rest of the culture by a gentle shakeoff step.

This procedure has to be optimized for each cell line to prevent leaving too many mitotic cells bound to the tissue culture flask or detaching too many interphase cells. Chromosomes isolated from nonadherent cultures, such as lymphoblasts, contain a higher fraction of interphase nuclei than do monolayer cultures. This presents a problem because of the higher number of nuclei that must be discriminated against when sorting. We require sorted chromosome fractions to contain less than one cell nucleus per 10,000 chromosomes. Cells suspended in growth medium at a concentration of 7×10^6 to 20×10^6 mitotic cells per ml. are gently centrifuged (250g for 10 minutes) to form a loose pellet. The growth medium is removed and replaced with a swelling buffer (40 mM to 75 mM KCL). The pellet is resuspended in the hypotonic buffer and allowed to stand (either at room temperature or at 4° C depending on success with a given cell line) for about 30 minutes. The cells are centrifuged again, the swelling buffer removed, chromosome isolation buffer added and the cell pellet resuspended. This is followed by mechanical shear using either vortexing or by shearing through a bent hypodermic needle. As the mitotic cells are sheared and the chromosomes are released, their structure is stabilized by the isolation buffer.

An ideal chromosome preparation would have the following properties. Only chromosomes and no cellular debris or interphase nuclei would be present in the suspension, every mitotic cell would have its chromosomes completely disassociated and the chromosomes would be monodispersed, the procedure would not degrade the DNA molecule within each chromosome, and finally the chromosome concentration would be at least 3×10^8 chromosomes/ml. This concentration allows flow sorting to approach the maximum efficiency for high speed sorting.

Typical chromosome preparations contain intact mitotic cells, and clumps of chromosomes, cell nuclei, and clumps of cells as shown in Figure 1. The representation of a single chromosome type can range from 0.5% to 4% of the fluorescent particles in suspension. The level of representation is highly dependent on careful optimization of all previous steps. There are also particles present in the suspension which are unmeasurable, such as cell membrane fragments, which do not stain with the fluorescent DNA dyes used to label the chromosomes. For most applications, the chromosomes are stained with chromomycin A3 and Hoechst 33258. chromomycin-A3 (CA3) preferentially binds to GC and Hoechst 33258 (Ho) to AT of the DNA (8). For some applications single fluorescent dyes such as propidium iodide or ethidium bromide are used to label total DNA content (4,8).

It is worth noting that some DNA stains are toxic compounds and some are suspected carcinogens. Buffers used for isolating chromosomes may also contain toxic chemicals.

Flow Cytometry

Flow cytometric sorting of chromosomes is singularly suited to the task of isolating microgram amounts of DNA of a specific chromosome type. Other separation methods are more rapid and provide larger amounts of DNA, but do not have the resolving power and therefore the purity of flow cytometric sorting. A flow cytometer measures the fluorescence of each particle as it traverses the sensing region of the instrument. If the particles are uniformly suspended, the rate of analysis appears constant, although the actual probability of arrival in the sensing region obeys Poisson statistics. The remainder of this section describes the Los Alamos High Speed Sorter.

Figure 2 illustrates the concept of hydrodynamic focusing of the sample suspension within the laminar flow field of an outer (or sheath) fluid. As the concentrated suspension of chromosomes enters the sheath fluid at the sample inlet tube, both the sample and sheath fluid undergo acceleration through a tapered region of the flow-chamber. The fluid enters the square capillary flow-chamber (250 μm by 250 μm inside cross sectional area, by 6 mm. long) and due to laminar flow it follows a constant velocity flow path along the axis of the flow-chamber. Hydrodynamic focusing in the tapered region confines the sample stream to a very small diameter (3 to 10μm) along a central flow line.

High power argon-ion lasers, Coherent Inc. models CR-12 and Innova 200-25/5, are used as excitation sources for the stained chromosomes. Their output beams (457.9 nm @ 1.0W for the CR-12 and 363.8 nm @ 1.0W for the Innova 200-25/5) are apertured to give TEMoo Gaussian profiles. These output powers require laser beam containment. The laser beam paths and the flow-chamber are enclosed to prevent direct or indirect laser beam exposure per ANSI standard Z-236.1 (9). The 457.9 nm beam is optically modified by a beam expander-collimator. The laser beams are brought in from opposite sides of the flow chamber (10) and are focused by crossed cylindrical lens sets to form elliptically shaped beam spots (approximately 20 μm vertical by 150 μm horizontal) on the sample stream. The focal point for the 363.8 nm beam is 50 micrometers upstream from the 457.9 nm beam (see Figure 2). A chromosome will pass through the focus of the 363.8 nm laser beam and approximately 12 microseconds later through the focus of the 457.9 nm laser beam. The 363.8 nm beam excites the Hoechst 33258 (Ho) molecules and the 457.9 nm beam will excite the chromomycin-A3 (CA3) molecules. Energy transfer from the excited state of a Ho molecule to a CA3 molecule occurs when these two dye molecules are bound in close proximity to each other on the DNA. Energy transfer can occur between energy levels in this dye combination due to the overlap of the emission spectra of Ho with the excitation spectra of CA3. In fact many of the excited Ho molecules transfer their energy to nearby CA3 molecules, thus the Ho

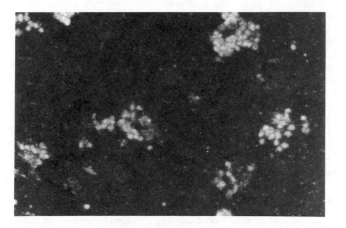

Figure 1. Microphotograph (250X) showing a typical chromosome preparation. Note the high number of large aggregates of nuclei in this sample.

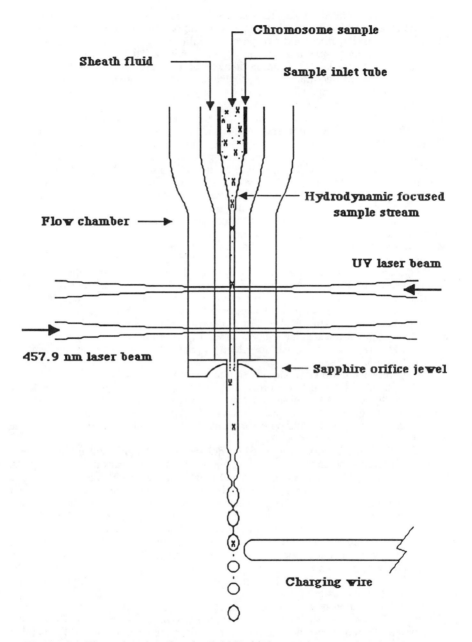

Figure 2. Flow-chamber for the LANL high speed sorter showing sample stream hydrodynamic focusing, laser beams entering from opposite sides of the flow-chamber, the synthetic sapphire cup jewel (with ~80 μm diameter orifice), and droplet formation along the exit jet. The charging wire is shown positioned near the break off droplet.

fluorescence is indirectly measured as CA3 fluorescence (8). The fluorescent emission is detected by an optical system consisting of a microscope objective, spatial filters, laser blocking filters and two photomultiplier tubes. The collection lens is a long working distance 40X microscope objective with a 0.6 NA. The spatial filter consists of two pinholes (each 1 mm in diameter and centered 2 mm apart) located in the image plane of the microscope objective. The pinholes define the spatial separation of the two optical paths. Two long-pass filters, 500 LP and 520 LP, eliminate the scattered laser light in the UV and 457.9 nm detector channels respectively. Two photomultiplier tubes are used, an RCA 4526 for UV excited energy transferred Ho detection and an EMI 9828B for detecting the CA3 emission. The signals from the photomultiplier tubes are amplified and actively integrated (11). The signal generated by a chromosome passing through the UV laser beam is delayed using a delay amplifier (EG&G ORTEC, Model 427A) to bring it into time register with the second laser beam. The two signals are then split and sent to parallel processing paths in the Los Alamos cell data analysis system (LACEL) (12). The first, a voltage comparator circuit, is an electronic rectangular window which analyzes the Ho and CA3 fluorescence intensities for each chromosome passing through the laser beams. If the intensities of the chromosome emission fall within those of the window, a sort command is issued. The decision time for this circuit is very short, nominally 5 microseconds, allowing up to 200,000 sort decisions per second. In parallel, the analog signals are also sent to two ADC modules (8 bit) which digitize the data for each chromosome. An LSI 11/83 computer (Digital Equipment Corporation) records the digitized values and stores them in a list mode data file. The data can be presented as a real time dot plot display, or the file can be processed to give one or two parameter histograms (13). From the dot plot display, the operator can interactively draw a rectangular window around the chromosome data of interest on the computer graphics screen, and download its parametric values to the sort window comparator circuit.

The sort decision (or command) is sent to a sort processing module which serves several functions. First, the sort processing module makes a pure sort decision by inspecting a time window before and after the sort command arrives from the sort comparator, to see if any additional particles not meeting the intensity values of the sort window are present. This pure sort function can be adjusted to inspect a time window from 2 to 8 drop periods long.

Secondly, the sort processing module delays the sort command for the time it takes for the chromosome to arrive at the point where the jet breaks into free droplets. For the High Speed Sorter this delay is usually 350 microseconds. A droplet period is one cycle of the ultrasonic drop generator (142.6 kHz., sheath pressure of 107 psi) or nominally 7.0 microseconds. The droplet period can be shortened (i.e. the frequency can be increased) by also increasing the sheath pressure. At a pressure of 200 psi the droplet period can be as short as 5.0 microseconds (f=200.0 kHz.). However, at 200 psi up to 80% of the sorted chromosomes adhere to the walls of the collection tube. A significant chromosome loss occurs due to this adherence. Purification of DNA from the adherent sorted chromosomes results in degradation of the molecular weight of the DNA. The present operation at 107 psi results in good recovery and high molecular weight (>1 MBase average) DNA.

The third function of the sort processing module is to drive the drop producing transducer. The transducer is an annular piezo-electric (PZT) element whose outer circumference is mounted rigidly to the instrument frame (nonmobile) while its inner circumference supports the flow-chamber. The sort processing module has a high voltage amplifier to provide the PZT drive from a signal generator (usually a sine wave source). The high voltage sine wave is applied to the PZT, causing it to distort axially up and down. The PZT motion superimposes a periodic velocity fluctuation on the jet as it exits the orifice. Due to surface tension effects, this perturbation of the jet diameter (or bunching of the fluid) causes the jet to break into drops at a very precise distance from the orifice. The distance can be changed by adjusting the amplitude of the signal driving the piezotransducer, changing the signal frequency, or changing the jet velocity. The purpose of delaying the sort command is to have the delay time end just as the chromosome arrives at the next droplet to be separated. As the droplet forms around the chromosome to be sorted, a high voltage (300 V) pulse is generated in response to the delayed sort command. This high voltage pulse is applied to a Pt-Ir wire electrode (400 μm diameter) placed 50 μm from the droplet(see Figure 2). The drop is connected to the jet by a fluid filament which is rapidly thinning to break off. The jet is electrically at ground potential (the sheath fluid is an ionic conductor). Thus a charge is induced on the surface of the drop, and results in a net electrostatic charge being carried away by the drop as it breaks free of the stream. The drops enter an electrostatic field. Depending on the polarity of their charge the drops will be deflected left or right as they fall through the electric field. Uncharged drops are collected by a waste tube. The high precision of the droplet formation, droplet charging, and finally the droplet trajectory, is the basis for ink jet printing (*14*). As applied to flow cytometry (*15*), its precision is dependent on holding the break off parameters (reference frequency, drop drive amplitude, and jet velocity) constant. One major perturbing factor is the presence of large particles such as a large cell, clumps of nuclei, or clumps of chromosomes passing through the orifice. As the fluid is accelerating, the presence of such particles will cause a perturbation in the pressure gradient in the orifice. This will result in a change in the dynamics of droplet formation and will cause the droplet break off point to move up the jet in an unpredictable manner (*16*). The presence of a large body in the orifice or in the connecting fluid filament at the moment of charging, can also reduce the electrostatic charge (by increasing the resistivity of the path to ground). Recent developments in monitoring the break off point have allowed us to detect break off point changes that result from clumps passing through the orifice (*17*). The arrival time of the disturbance is generally the same as the sort command delay time. The perturbation in the break off point can last for 500 microseconds or more (though some 100 and 200 microsecond disturbances occurred) as seen by triggering an oscilloscope when a clump of nuclei passed through the laser focal points. This indicates that the presence of a large mass in the orifice causes a hydrodynamic disturbance which affects the formation of those droplets following it, as it travels down the jet to the break off point. The length of time the break off distance is disturbed from equilibrium is directly related to the size and shape of the mass causing it. It is clear this represents an undesirable event to have to analyze and discriminate against. One cannot accurately predict where the break off point truly is when these events occur. We have been able to minimize this effect by use of the wire charging electrode and by single drop sorting. The small air gap between the wire and the equilibrium break off point creates an intense electric field. As the break off moves up the jet in response to a clump (*16*), the capacitive coupling is drastically reduced, resulting in very little charge being deposited on incorrect droplets. Single drop sorting, rather than the more conventional 2 or 3 drop charging, therefore results in the minimum system deadtime, the maximum purity and the highest concentration of sorted chromosomes. High speed sorting is based on increasing the chromosome throughput or analysis rate in concert with increasing the droplet frequency. Conventional sorters with droplet frequencies of 4×10^4 droplets per second and throughput rates of 2×10^3 chromosomes per second have an average spacing of 20 droplets between chromosomes. Increasing the analysis rate to 2×10^4 chromosomes per second reduces the average spacing to 2 droplets between chromosomes. With average events spaced this close together, coincidence problems would make the recovery of pure sorted populations very inefficient, even with single droplet sorting. Therefore to increase the effective sorting rate the droplet frequency must be greatly increased.

In the Los Alamos high speed sorter we are using frequencies of approximately 1.5×10^5 droplets per second, which results in an average spacing of seven and half droplets between chromosomes at rates of 2×10^4 chromosomes per second. A better way to visualize the crowding of events is in terms of the Possion statistics of the occurrence of chromosomes in droplets. With this statistical analysis we can examine the probability for chromosome pile up, *i.e.* the coincidence of two or more chromosomes in the same droplet. This analysis is shown in Table I. Note, that the pile up of chromosomes in droplets at the 2×10^3 chromosomes/sec sample rates correspond to about 23% and 6% of the droplets containing chromosomes at droplet frequencies of 40 kHz and 150 kHz respectively. From a sorting stand point additional losses would occur using more than one droplet sorting and/or a pure sort decision circuit. Coincidences effects and sort purity are discussed further by Miller and McCutcheon (*18*).

TABLE I. Distribution of Chromosomes in Droplets

Droplet Frequency: (droplets/sec)	40,000		150,000	
Chromosome Rate: (chromosomes/sec)	2,000	20,000	2,000	20,000
Average occupancy: (chromosomes/droplet)	0.05	0.50	0.01	0.13
Fraction* of droplets containing:				
0	0.951	0.606	0.990	0.878
1	0.048	0.303	0.010	0.114
2 or more chromosomes	0.001	0.091	0.000	0.008

* Calculated from the Poisson distribution formula: $P(s;m) = e^{-m}m^s/s!$,
where: m = mean occupancy per droplet and s = number of chromosomes per droplet.

The rate at which the high speed sorter can analyze and processes chromosomes (20 kHz.) and the frequency with which it makes droplets (142 kHz. to 220 kHz.) have determined our present sorting strategy. Chromosome sample concentration is the primary factor in determining sort rate. The presence of clumps can reduce the rate considerably due to the effects clumps have on purity. Table II lists purity and sort rates routinely achieved for several experiments. In the first row, a mixture of different diameter fluorescent microspheres was used to mimic the effects of a low concentration of desired chromosomes in the presence of larger chromosomes and nuclei. The purity for sorting 0.9 μm diameter microspheres was assayed by sorting 5×10^5 microspheres onto a membrane filter. The contaminating microspheres were then counted using a fluorescence microscope. Sixty five large microspheres were found in the sort of half a million small microspheres.

The remainder of Table II presents the chromosome sort purity achieved for several different cell lines while analyzing at different rates. Sort purity was assayed by taking aliquots of the sorted chromosomes (50 microliters which represents approximately 50,000 chromosomes) and processing them by fluorescence in-situ hybridization. The hybridization was carried out using either chromosome specific repeat probes or total human DNA probes (*19*). The effect of analysis rate on sort purity is clearly seen in GM2184A. This cell line has a normal euploid karyotype as determined by standard karyotype analysis. Using this cell line, a number of experiments were carried out to isolate chromosome 16, one of the E group chromosomes. Because the fluorescent intensity of chromosome 16 is not very different from chromosome 15 or 17, the increase in impurity with higher analysis rate is due to greater overlap in the statistical description of the fluorescence from each population. Therefore, with higher analysis rates, one reaches a limit where an attempt to increase the rate (by increasing the sample flow rate) results in degraded measurement resolution which in turn results in degraded sort purity. To improve the statistical discrimination of a human chromosome from the rest of the human karyotype, hybrid cell lines are employed. The hybrid cell is a cell fusion product which may contain elements of more than one genome. These cells are cloned and karyotyped to determine which chromosome types are retained by the hybrid cell. Those cells which retain a single or only a few human chromosomes become candidates for flow karyotyping and sorting. Hybrid cell lines are screened by flow analysis for the presence of desired chromosomes that are well resolved from the chromosomes of the host cell line. Several of these hybrid cell lines were used for library construction. Table II illustrates that when a chromosome type is better resolved, the purity of the sorted fraction becomes less dependent on the analysis rate. Here the effects of small fractured pieces of chromosomes (below the detection level),

Table II. Sort Purity Studies for High Speed Sorter

Sample	Population Sorted	Analysis Rate (sec^{-1})	Sort Rate (sec^{-1})	Purity[*] (%)
μ-sphere Mixture	0.9 μm (1%)	20,000	200	99.9[+]
GM2184A # 16 (Human Lymphoblasts)		2,500 6,000 10,000	60 90 150	93 85 77
CY-18 # 16 (Mouse-Human Hybrid)		2,000 12,000	20 70	87 87
UV20 HL21-27 # 4 (Hamster-Human Hybrid)		5,000 18,000	100 200+	98 91
Q826-20 # 5 (Hamster-Human Hybrid)		10,000	100	94
WAV-17 # 21 (Mouse-Human Hybrid)		20,000	300[+]	89

* By *in-situ* hybridization using chromosome specific probes (2).

clumps of nuclei, undispersed metaphase cells, and clumps of chromosomes result in decreased sort purity. To further improve sort purity one must also address the problem of droplet generation instability as described above. We are pursuing methods that remove large events (which disrupt droplet generation), remove small debris, and which increase the number representation of the chromosome one wishes to sort from 1% to 30% or more.

Prefractionating Chromosome Preps Before Flow Sorting

Velocity Sedimentation. This technique has been extensively pursued by A. Tulp, who has developed a sedimentation chamber which has features optimized for separating chromosome types (20). The velocity sedimentation chamber, Figure 3, has a large surface area (50 cm^2). A standard 5 ml. sample forms a thin 1 mm layer on top of the sucrose gradient. The gradient can be either 5% to 15% or 5% to 30% depending on the application. Streaming caused by nuclei or clumps is minimized due to the sample layer being so thin. The chamber holds 100 ml. of gradient solution in which to separate the chromosome suspension. The top and bottom of the chamber are specially machined cones which are fitted with flow deflectors at the inlet/outlet ports. The flow deflectors aid in the smooth filling of the chamber, with loading the chromosome sample onto the sucrose gradient and in the collection of fractions by backfilling the chamber from the bottom port. When using the chamber in a centrifuge, an anti-vortex cross is inserted to prevent circulatory flow during acceleration phases. The method for using the chamber calls for filling the chamber with a sucrose cushion layer (15% or 30% depending on the steepness of the gradient) through the bottom port. A peristaltic pump is used to smoothly fill the chamber. The gradient is introduced through the top inlet as the cushion layer is released through the bottom port. When the gradient has been loaded (a five minute procedure), 3 to 5 ml. of sample is loaded through the top inlet by removing more of the cushion layer, again about a five minute step. Finally a buffer layer (28 ml. of chromosome isolation buffer (CIB) without sucrose) is layered on top of the sample as more cushion fluid is removed through the bottom port. Centrifugation at 52g for 50 minutes at 4° C is adequate to separate the human karyotype in the 100 ml. gradient solution (21). Alternatively, the chamber can be left at unit gravity for 18 hr. at 4° C (22). Variations on the procedure such as a two part gradient, 4° C vs. room temperature, and amount of g-force used, imply that this is a flexible method that can be optimized for a specific experimental requirement. Unfortunately there are a several drawbacks. The procedure works best if chromosome size (degree of condensation) does not vary for a particular chromosome type. The degree of condensation will influence the chromosome length and in turn, how uniformly it will sediment in the gradient. Tulp reports that cell culture synchronization will minimize variation in the degree of condensation in mitotic chromosomes (22). A second major difficulty arises from the dilution of the chromosome concentration. After fractionating the gradient, a particular chromosome type is often present in several 5 ml. fractions. Thus the 5 ml. of initial sample was diluted to 15 ml. of collected gradient material. While the representation of a chromosome type will be enhanced on a percentage basis in a fraction, the overall concentration is reduced. Ideally, one would prefer to reconcentrate the chromosomes in the fractions. Tulp describes a concentrating step where he either centrifuged the collected fraction (3000 rpm for 45 minutes, no g force given) or allowed the chromosomes to sediment to the bottom of the sucrose fraction while storing it for two weeks at 4° C (22). Collard indicates that two parameter analysis resulted in event rates of 200 chromosomes per second following velocity sedimentation (21). Our experience has confirmed a chromosome loss of up to 50%, and a dilution, which together give an analysis rate of 200 events per second. In the same paper, Collard presents single parameter data where the event

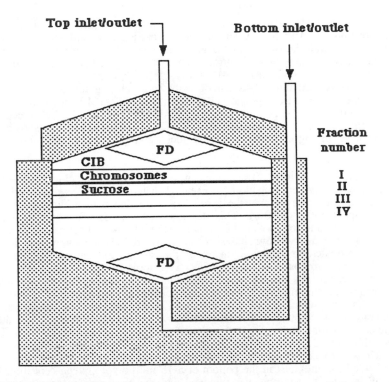

Figure 3. Diagram showing the Tulp chamber, adapted from Collard *et. al.* (*21*). The top and bottom loading ports are indicated by the arrows. The bottom of the chamber is filled with a sucrose gradient. The chromosome preparation is layered on top of the sucrose gradient. Chromosome isolation buffer (CIB) is layered on top of the chromosome preparation. The interface between the chromosome preparation and the sucrose gradient is indicated by the heavy line. The fraction numbers are indicated along the side of the chamber. Note the top and bottom flow deflectors (FD) which minimize mixing during filling and draining. Not shown is an anti-vortex cross used to minimize fluid rotation during acceleration phases.

rate was 1000 chromosomes per second (no sample volumetric flow rate is given). Fukushige et. al. (23) have used velocity sedimentation to separate chromosomes on sucrose gradients prior to flow cytometric sorting. They report analyzing chromosomes (single parameter) at rates of 1000 to 2000 chromosomes per second following velocity sedimentation. Although no details are provided, it would appear that in order to achieve these high analysis rates, the chromosomes in each fraction were reconcentrated.

Isopycnic Centrifugation.   Isopycnic separation is based on using buoyant density banding of particles. Centrifugation is carried out until all particles have reached their equilibrium point in the gradient. Metaphase chromosomes have an apparent buoyant density of 1.288 g/ml while nuclei buoy at 1.249 g/ml. as observed by banding in a Nycodenz (24), {N,N'-bis(2,3-dihydroxypropyl)-5-[N-(2,3-dihydroxypropyl)acetamido]-2,4,6-triiodo-isophthalamide}, gradient (25). We have used a related iodinated compound called Metrizamide (24), {2-[3-acetamido-5-(N-methylacetamido)-2,4,6-triiodobenzamido] -2-deoxy-D-glucose}, for recovering chromosomes suspended in sucrose solutions (26). This method has shown some promise of separating chromosomes from nuclei and in concentrating chromosome preparations.

Concentrating Chromosomes.   We were able to recover chromosomes from sucrose fractions by centrifugation onto several different media as described below. The first approach was to layer the chromosome fraction onto a high concentration sucrose solution (40% or 60%). This step gradient was centrifuged at 500g for 20 minutes. The chromosomes would band at the step interface. The main disadvantage to this approach was that flow analysis of the recovered chromosomes was degraded due to a mismatch in refractive index of the sheath and sample solutions.   In recovering the chromosomes from the interface, some of the 40% or 60% sucrose below the interface was also recovered thereby increasing the sucrose concentration of the chromosome suspension. To verify this conclusion, highly uniform fluorescent latex beads (Polyscience, 2.02 μm diameter) were suspended in similar sucrose concentrations. The optical properties in the flow chamber were changed such that the coefficient of variation increased significantly for each parameter (U.V. had been 1.6% and became 3.0%, 457.9 nm had been 1.0% and has become 2.5%). A second approach was to centrifuge the chromosomes onto a high density agarose layer (2.50% w/v in saline) cast in the bottom of a 15 ml. polypropylene centrifuge tube. Most of the sucrose solution was removed and the chromosomes resuspended by pipetting the remaining solution over the agarose surface. Chromosome concentration was increased two to five-fold but 10% to 50% of the total number of chromosomes were retained in the agarose unless the surface was rinsed very thoroughly. Excess rinsing would dislodge pieces of agarose which had to be filtered out with 37 μm nylon mesh. The procedure also suffered from the occasional dislodging of the entire agarose plug while centrifuging.   The final procedure was to layer the sucrose fractions onto a Metrizamide solution (20% w/v in saline) and recover the chromosomes that collected in the Metrizamide layer and at the interface.   This resulted in a 10X concentration but again the refractive index of the Metrizamide solution yielded poor optical resolution when analyzing the chromosomes by flow cytometry (26).

Nuclei Sedimentation.   A novel use was made of the Tulp sedimentation chamber in an attempt to remove only the large clumps of nuclei from a chromosome preparation. The chamber was filled with 5% (w/v in polyamine buffer, 4°C) sucrose solution. Five ml. of stained chromosome sample were introduced through the top port while slowly draining the 5% sucrose solution out of the bottom port. The fluid level was lowered to the point where the stained chromosome suspension could be viewed at the top of the cylindrical portion of the sedimentation chamber. The chamber was observed for the next 5 minutes, and large clumps could be seen rapidly sedimenting toward the bottom of the chamber. Sucrose was slowly reintroduced through the bottom port, and the chromosome sample recovered as Fraction I (the first 5 ml fraction pumped out of the top of the chamber). Fraction II was the sucrose layer underneath the chromosome sample (5 ml). Fractions III and IV were the next 5 ml fractions in succession. Photographs of these fractions show that large clumps were effectively removed from Fraction I (see Figure 4). Flow cytometric data indicates that very few human 21 chromosomes were lost, see Table III. The sort rate for chromosome 21 is higher than the control which was not fractioned. A possible explanation for the apparent increase in concentration may be due to the elimination of the large clumps. Under normal operating conditions these clumps settle out of suspension in the 1.0 ml. syringe used for sample injection into the flow cytometer. The clumps sediment through the chromosome suspension under 1g acceleration, and by streaming action carry many chromosomes out of suspension. Indeed, one often sees a decrease in both the chromosome count rate and the sort rate over time.

Table III Nuclei Sedimentation Analysis

Fraction Number	Analysis Rate (sec^{-1})	Nuclei Rate (sec^{-1})	21 Sort Rate (sec^{-1})
Control	20,000	67	286
I.	17,000	71	362
II.	< 1,000	39	52
III.	< 500	4	20
IV.	< 250	<1	6

Conclusions

Flow sorting has a long and successful track record for chromosome separation (27,28). The molecular weight of recovered DNA at the present time is sufficient for bacteriophage, cosmid, and yeast artificial chromosome cloning systems (>1 MBase). As new larger insert vectors are pursued, the amount of DNA needed to successfully create a library also increases. Therefore the effort required for sorting library quantities of DNA has significantly increased during the past six years of the National Laboratory Gene Library program. The application of velocity sedimentation fractionating is being pursued as a technical advance to improve sorting rates. Because of the dilution effects however, we are evaluating methods for concentrating the fractionated samples. Recent efforts show that nuclei can be differentially sedimented out of the preparations, which is a positive step toward faster and higher purity sorting.

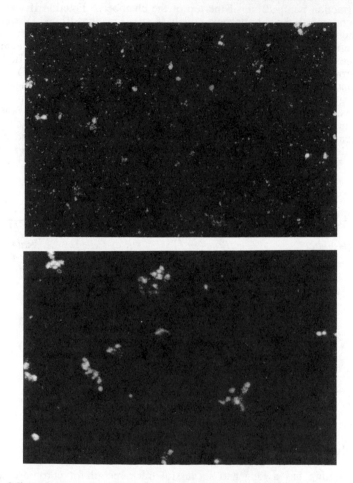

Figure 4. Microphotographs of the first four fractions recovered from the Tulp chamber after layering the chromosome preparation shown in Figure 1 onto a 5% sucrose solution for 5 minutes. Panels A, B, C, and D are fractions I, II, III, and IV respectively. Fraction I has a high concentration of chromosomes and some individual nuclei, but very few large clumps. Fractions II through IV show large clumps but decreasing numbers of chromosomes further in the gradient.

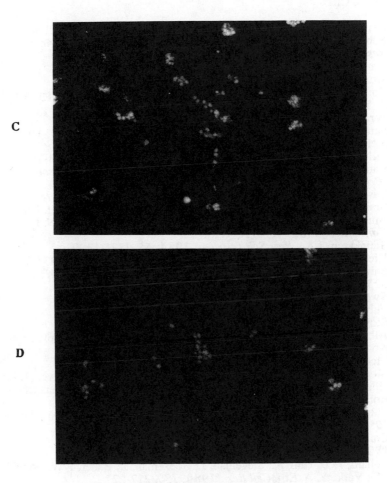

Figure 4. Continued

Acknowledgments

The authors wish to thank B. Hoffman, B.F. Bentley, and R.E. Roemer, for their technical assistance, and J. H. Jett for helpful discussions. This work has been supported by the U.S. Department of Energy and the National Flow Cytometry Resource (NIH grant RR01315).

Literature Cited

1. Griffith, J.K.; Cram, L.S.; Crawford, B.D.; Jackson, P.J.; Schilling, J., Schimke, R.T.; Walters, R.A.; Wilder, M.E.; Jett, J.H. *Nucleic Acids Research,* **1984**, *12*, 4019-4034.
2. Van Dilla, M.A.; Deaven, L.L.;*Cytometry,* **1990**, *11*, 208-218.
3. American Type Culture Collection, Rockville, MD.
4. Bartholdi, M.F.; Ray, F.A.; Cram, L.S.; Kraemer, P.M. *Cytometry,* **1984**, *5*, 534-538.
5. Wray, W.; Stubblefield, E. *Exp. Cell Res.,* **1970**, *59*, 469-478.
6. van den Engh, G.; Trask, B.; Cram, S.; Bartholdi, M. *Cytometry,* **1984**, *5*, 108-117.
7. Sillar, R.; Young, B.D. *J. Histochem. Cytochem.,* **1981**, *29*, 74-78.
8. Langlios, R.G.; Carrano, A.V.; Gray, J.W.; Van Dilla, M.A.*Chromosoma (Berl.)* **1980**, *77*, 229-251.
9. Laser Safety Committee, Laser Institute of America; In *Laser Safety Guide*; Smith, J.F. Ed; Sixth Ed.; Laser Institute of America: Toledo, OH, **1987**.
10. Burchiel, S. W.; Martin, J. C.; Amai, K.; Ferrone, S.; Warner, N. L. *Cancer Res.* **1982**, *42*, 4110-4115.
11. Steinkamp, J.A., *Rev. Sci. Instrum.,* **1984**, *55*, 1375-1400.
12. Hiebert, R.D.; Jett, J.H.; Salzman, G.S., *Cytometry,* **1981**, *1*, 337-341.
13. Salzman, G.S.; Wilkins, S.F.; Whitfill, J.A.,*Cytometry,* **1981**, *1*, 325-336.
14. Sweet, R.G. *Rev. Sci. Instrum.,* **1965**, *36*, 131
15. Fulwyler, M.J. *Science,* **1965**, *150*, 910-913.
16. Stoval, R.T. *J. Histochem. Cytochem.,* **1977**, *25*, 813-820.
17. Albright, K.L.; Martin, J.C.; Cram, L.S. *Cytometry,* **1990**, Supplement No.4, 435A.
18 McCutcheon; M.J., Miller, R.G. *Cytometry,* **1982**, *2*, 219 .
19. Moyzis, R.K.; Albright, K.L.; Bartholdi, M.F.; Cram, L.S.; Deaven, L.L.; Hildebrand, C.E.; Joste, N.E.; Longmire, J.L.; Meyne, J.; Schwarzacher-Robinson, T. *Chromosoma (Berl.),* **1987**, *95*, 375-386.
20. Tulp, A.; Collard, J.G.; Hart, A.A.M.; Aten, J.A. *Anal. Biochem.,* **1980**, *105*, 246-256.
21. Collard, J.G.; Philippus, E.; Tulp, A.; Lebo, R.V.; Gray, J.W. *Cytometry,* **1984**, *5*, 9-19.
22. Collard, J.G.; Tulp, A.; Stegeman, J.; Boezeman, J.; Bauer, F.W.; Jongkind; J.F.; Verkerk, A. *Exp. Cell Res.,* **1980**, *130*, 217-227.
23. Fukushige, S.; Murotsu, T.; Matsubara, K. *Biochem. and Biophy. Res. Comm.,* **1986**, *134*, 477-483.
24. Sigma Chemical Company, St. Louis, Mo.
25. Gollin, S.M.; Wray, W. *Exp. Cell Res.,* **1984**, *152*, 204-211.
26. Hoffman, B. Private communication.
27. Griffith, J.K., Cram, L.S.,Crawford, B.D., Jackson P.J., Schilling J., Schimke, R.T., Walters, R.A., Wilder, M.E.; Jett, J.H. *Nucleic Acids Res.,* **1984**, *12*, 4019-4034.
28. Gray, J.W.; Cram, L.S. In *Flow Cytometry and Sorting.* Melamed,M.R., Lindmo, T; Mendelsohn, M.L., Eds; Second Ed. Wiley-Liss: New York, NY, **1990**, 503-529.

RECEIVED March 15, 1991

SEDIMENTATION AND FLOW

Chapter 6

Separation of Cells by Sedimentation

Thomas G. Pretlow and Theresa P. Pretlow

Department of Pathology, Case Western Reserve University,
Cleveland, OH 44106

Isopycnic and velocity sedimentation have proved to be
broadly applicable techniques for the separation of
mammalian cells according to their physical properties,
i.e., density and effective diameter. Isopycnic sedimentation has been useful for the determination of the
densities of cells; however, isopycnic sedimentation
has been less widely applicable than velocity sedimentation for the purification of most kinds of mammalian
cells. There are four techniques for velocity sedimentation that have been used by many: elutriation, sedimentation at unit gravity, sedimentation in an isokinetic gradient of Ficoll in tissue culture medium, and
sedimentation in a reorienting gradient zonal rotor.
All four of these techniques for velocity sedimentation
are used with shallow (gm/ml/cm) gradients and are capable of high-resolution separations of cells based on
their rates of sedimentation. The kinds of data required for the evaluation of cell separations are discussed. Because of the broad scope of this subject, we
have attempted to supply references that will provide
specific details that will not fit in this brief review.

Velocity sedimentation and isopycnic centrifugation in continuous
gradients are widely used techniques for the preparative and analytical separation of cells. Historically, the first serious efforts
to purify cells by means of sedimentation were in the laboratory of
Lindahl (1). Lindahl has presented a critical analysis of the technique that he invented (2) and called counter-streaming centrifugation. This technology was transferred to other laboratories slowly
because of the requirement for relatively expensive equipment that
was not readily available commercially; however, the Beckman Instrument Company has manufactured a slightly modified form of Lindahl's
instrument and has called it an "elutriator." In the 1960s and

0097–6156/91/0464–0090$06.00/0

1970s, other approaches to cell separation by sedimentation were developed rapidly. These include sedimentation at unit gravity, sedimentation in an isokinetic gradient of Ficoll in tissue culture medium, cell separation in a reorienting gradient zonal rotor, and isopycnic sedimentation in continuous gradients.

Theory

In the design of experiments for the separation of cells, perhaps the most useful and most commonly neglected fact is that there are only two physical properties of cells that determine their rates of sedimentation: diameter and density. If investigators were more widely cognizant of this fact, methods for the separation of cells could be designed much more efficiently; and the resultant purifications of cells would be of much higher quality. The sedimentation of cells is governed by the Stokes equation which is given below.

$$v = dr/dt = [2a^2(\rho_p - \rho_m)\omega^2 r]/9\eta$$

In this equation, r represents the distance of the cell from the center of revolution; t, time; dr/dt, the velocity of the cell; a, the radius of the cell; ρ_p, the density of the cell; ρ_m, the density of medium in which the cell is suspended at the location of the cell; ω, the angular velocity of the centrifuge; and η, the viscosity of the medium in which the cell is suspended at the location of the cell.

For cells that are not spherical, we can determine an effective value for the radius, i.e., a in the Stokes equation. We should emphasize the fact that the integration of the Stokes equation can be used effectively for cells such as neurons and osteoclasts that are not spherical even in suspension. One can use experimentally determined velocities of sedimentation and other measured or known parameters in the Stokes equation to solve for what we (3) have termed the "effective diameter." The effective diameter cannot be proved to describe anything more about a particular cell than the factor (a) that describes the behavior of that cell during sedimentation; however, with a perfectly circular error and great practical utility, this value for "effective diameter" can be used with the integrated Stokes equation for the simulation of the sedimentation behavior of the cell under consideration and for the design of conditions and gradients that will permit the optimal purification of that cell.

Boone et al. (4) first proposed that the integration of the Stokes equation might provide a useful approach to the design of experiments in the separation of cells by velocity sedimentation. Subsequently, we (5) demonstrated for the first time that the Stokes equation, integrated by a trapezoidal computer integration, accurately describes cell separations as observed during gradient centrifugation in the laboratory. Both trapezoidal (5) and closed (6) integrations of the Stokes equation for the purification of cells have been described.

Isopycnic Centrifugation

In the "isopycnic centrifugation" of cells, one separates cells according to their buoyant densities and uses a sufficient product of centrifugal force and time, $\omega^2 rt$, to sediment cells to the locations in the gradient where the densities of the gradient and the cells are equal. From the Stokes equation it becomes apparent that, when the cells approach their densities in the gradient, their velocities approach zero; and additional time or centrifugal force will not change their positions in the gradient. Leif and Vinograd (7) were the first to realize the importance of continuous gradients for the isopycnic centrifugation of cells and were the first to separate cells by isopycnic centrifugation in continuous gradients. They used this technique for the purification of subpopulations of erythrocytes (7). The two groups that pioneered the application of isopycnic centrifugation in continuous gradients for the separation of cells were Leif's group and Shortman's group. Both of these investigators have reviewed cell separation by sedimentation (8,9). The most widely used potentially isotonic media for the isopycnic centrifugation of cells have been albumin (7,10), colloidal silica (11, 12), and colloidal silica coated with polyvinylpyrrolidone and given the name "Percoll" (13). In addition to these, many other media have been used for the isopycnic centrifugation of cells including Ficoll (4,5,14-24), Metrizamide or 2-(3-acetamido-5-N-methylacetamido-2,4,6-triiodobenzamido)-2-deoxyglucose (25-27), and many others (3,28,29). Many of these media have specific disadvantages. For example, Ficoll is more viscous than most gradient media and requires very high centrifugal forces for isopycnic centrifugation. Albumin and Percoll are probably the best media for the isopycnic sedimentation of cells. Percoll offers the advantage that it is commercially readily available in a form that requires little further preparation. Note that our understanding of the osmolarity of Percoll is different in recent years (13,30).

     In nature, most recognizable single types of cells exhibit broad and overlapping ranges of density (3). As a result of this broad range of densities observed for most mammalian cells, isopycnic sedimentation has been relatively limited in its applications for the purification of cells. It has been particularly useful for the separation of red blood cells and subpopulations of red blood cells (7), various classes of mast cells (16), cardiac myocytes (19), pancreatic acinar cells (23), parotid acinar cells (31), and rat proximal tubule cells (32).

Velocity Sedimentation

"Velocity" or "rate-zonal" sedimentation refers to the use of experimental conditions under which sedimentation is stopped before cells arrive at their isopycnic densities in the media in which they are sedimented. Most well designed systems for velocity sedimentation employ gradients in which the actual densities of the cells being separated are not even represented. Cells of equal density but different diameters can be separated by velocity sedimentation. Cells of equal diameter but different densities can be separated independently by velocity or by isopycnic sedimentation. Since cells are

separated during isopycnic centrifugation based on differences in
the densities of cells and during velocity sedimentation based on
differences in rates of sedimentation, a property that is dependent
upon both diameter and density, it is occasionally useful to employ
a two-step procedure in which velocity sedimentation and isopycnic
centrifugation are used. When this approach is taken, isopycnic
centrifugation is often most useful as the second step, since very
large sample volumes composed of fractions collected from velocity
sedimentation or from other procedures for purification can be layered directly over gradients for isopycnic centrifugation. Starting
samples for velocity sedimentation must be smaller and of low density. This two-step approach to the separation of cells by sedimentation, i.e., velocity sedimentation followed by isopycnic sedimentation, was first employed in 1970 (16,33) and has been used for
several different problems in the separation of cells since 1970.
     Velocity sedimentation in an isokinetic gradient of Ficoll in
tissue culture medium has been applicable for the separation of many
different kinds of cells. We have reviewed this method as used in
our laboratory and in the laboratories of others recently (34). The
uses of other methods for the velocity sedimentation of cells have
also been reviewed. These approaches have included elutriation (35,
36), sedimentation in a reorienting gradient zonal rotor (37-41),
and sedimentation at unit gravity (42-46).

Sedimentation at Unit Gravity

Sedimentation at unit gravity was first used by Mel and his collaborators (47-50). In its most widely used form, sedimentation at unit
gravity has been carried out with slight modifications of a technique described by Peterson and Evans (45). The theory of velocity
sedimentation as used at unit gravity has been described (46,51).
The method for separating cells at unit gravity has the advantage
that the required equipment is inexpensive. While this method has
been employed for the purification of large numbers of cells, it is
best suited for the purification of less than 10^8 cells. We (3) and
others (42-44,46) have reviewed applications of cell separation at
unit gravity previously.
     The time required for sedimentation of cells at unit gravity
can be shortened considerably with the use of a reorienting gradient
in a device invented by Bont and Hilgers (52). This apparatus is
available commercially as modified by Wells (53). The major disadvantage in the use of unit gravity sedimentation results from the
fact that only the earth's gravitational field is available. This
results in a slow separation that generally requires at least two
hours and may require sedimentation overnight. If the cells tend to
aggregate, as the majority of viable cells do, aggregates may form
as cells are together in the same zone of the gradient for hours.
If aggregates form, they sediment with larger effective diameters
(defined in 3) and are found in zones of the gradient which contain
cells that are larger than the monomers of the cells that form the
aggregate; i.e., aggregation will alter the sedimentation of cells
and generally cause a decreased purity and/or recovery of cells
after sedimentation. If aggregation results in aggregates that are
heterogeneous with respect to the kinds of cells that are incorpo-

rated as is generally the case, this precludes a high degree of
purification of the cells involved in these aggregates.

Isokinetic Sedimentation

Isokinetic sedimentation in a gradient of Ficoll in tissue culture
medium, like sedimentation at unit gravity, is best suited for the
purification of relatively small numbers of cells. For isokinetic
sedimentation, in addition to any refrigerated, swinging bucket centrifuge, only commonly available, inexpensive laboratory equipment
is required. One can separate up to approximately 35 million cells
per gradient with the technique that we have described (29,34,54,
55). It is possible to run three or four gradients so that the
range of cells to be purified may go as high as or slightly higher
than 10^8 cells. As a practical approach to the purification of
cells, isokinetic sedimentation has been employed for the purification of a wide range of types of cells in many different laboratories (34).

Elutriation

The most commonly employed approach to velocity sedimentation for
the purification of large numbers of cells is elutriation. There
have been many reviews of the purification of cells by elutriation
(2,35,56,57). Elutriation has been used extensively and successfully by many investigators experienced in the use of this technique
(36,58-63). It is our impression from discussions with many who use
and who have tried to use the elutriator that this technique has
resolution that closely approaches the resolution that one obtains
with sedimentation at unit gravity or sedimentation in the isokinetic gradient; however, it appears that the most effective use of
the elutriator requires considerably more experience and expertise
than the use of either of these other two methods for velocity sedimentation. Additional aspects of elutriation are detailed in the
next chapter.

Reorienting Gradient Zonal Rotor

A relatively simple method for the purification of approximately the
same number of cells as can be purified by elutriation is centrifugation in a reorienting gradient zonal rotor. Like elutriation, the
use of this technique requires a special centrifuge head. Both
techniques are capable of purifying 10^8-10^9 cells. Neither technique works well for very small numbers of cells. The use of the
reorienting gradient zonal rotor requires very little experience;
however, it has not been promoted effectively, and most of the
scientific community seems to be unaware of this approach. Separation of cells in the reorienting gradient zonal rotor has been reviewed and has been employed for the purification of many kinds of
cells (37-41,64).

Comparison of Isopycnic and Velocity Sedimentation

Comparisons of velocity and isopycnic sedimentation have been made
and discussed in detail previously (3,28,29,65). Isopycnic sedimentation is well suited for the purification of several kinds of cells
as described above; however, as a means of purifying cells, isopycnic sedimentation has more limitations than does velocity sedimentation. We are aware of only a few specific applications in which
morphologically recognizable kinds of cells can be better purified
by isopycnic sedimentation than by velocity sedimentation. These
applications include the purification of various kinds of mast cells
(16), the separation of red blood cells from some varieties of nucleated cells, and the purification of rat proximal tubule cells
(32). Interestingly, while one can purify rat proximal tubule cells
relatively effectively by isopycnic sedimentation (32), hamster
proximal tubule cells are more effectively purified by isokinetic
sedimentation than by isopycnic sedimentation (66). If one is able
to obtain approximately comparable purification by velocity sedimentation and by isopycnic sedimentation for a particular kind of cell,
velocity sedimentation offers the advantage that the cell will be
exposed to many-fold less centrifugal force when separated by velocity sedimentation than when separated by isopycnic sedimentation.
While certain cell functions have been shown to be capable of sur-
viving forces as high as 20,000g (67), most cells are severely
injured or killed by these very large centrifugal forces.

Discontinuous Gradients

Discontinuous gradients have been widely employed for the purification of blood cells. Red blood cells are present in human peripheral blood at approximately 500 to 1,000-fold higher concentrations than are nucleated cells. If one wishes to purify a particular kind of nucleated cell from peripheral blood, it is not feasible
to do this under ideal circumstances because of this large initial
ratio of red blood cells to nucleated cells. In order to get back a
useful number of nucleated cells, it is necessary to load a much
larger number of cells on a gradient than is optimal. As discussed
in detail previously (3,28) each gradient has a maximum capacity
known as a band capacity. When this is exceeded, nonideal sedimentation occurs; and generally lower degrees of purification are
achieved. Discontinuous gradients have been necessary for the separation of nucleated cells from erythrocytes because of this enormous
problem related to the ratio of red blood cells to nucleated cells.
Boyum developed methods that are commonly used for this approach
(68-74). It should be noted that Boyum pointed out that centrifugal
force is not required for the separations that he achieved. The
process is somewhat accelerated by the use of a centrifuge. Boyum
used a gradient medium which causes red blood cell agglutination,
and this is an important characteristic of the medium which Boyum
explored for blood cells. The limitations of this valuable method
have been discussed in detail (75).

The use of discontinuous gradients for most purposes is not advantageous. When one uses a discontinuous gradient, it is necessary
for the cells to pass through a succession of interfaces where they

are concentrated and often aggregate. Another disadvantage results
from the fact that cells cannot be maximally resolved in discontinuous gradients since the continuous range of densities and viscosities cannot be represented in discontinuous gradients. The absence
of gradients in the intervening distances between successive interfaces results in a loss of the stabilization of the column of fluid
that is the major advantage of continuous gradients. Thus, in the
absence of continuous gradients, several artifacts are enhanced as
has been discussed by many authors (3,8,9,28,76).

     For reasons that seem less than justified to us, some inves-
tigators have been attracted to the "sharp" bands that are obtained
at interfaces in discontinuous gradients. It is possible to combine
fractions of different kinds of cells separated in continuous gradients. It is impossible to separate the different kinds of cells
that would have been separated in a continuous gradient by centrifuging them against the barriers of increased density and viscosity
that are encountered at the interfaces in discontinuous gradients.
As stated by de Duve (76): "The discontinuous gradient is essentially a device for generating artificial bands. This may be a convenient way of compressing together for preparative purposes certain
segments of the distributions observed in continuous gradients. But
it is also a very dangerous procedure, in that it creates the illusion of clear-cut separation...Density-gradient centrifugation in a
continuous gradient is the analytical method 'par excellence'. It
lends itself to an entirely objective assessment of the frequency-
distribution curves of certain physical properties, such as density
or sedimentation coefficient, from which in turn other characteristics of a population, including its size distribution, can be
derived."

Design of Gradients

The design of gradients for isopycnic centrifugation has been discussed (3). Since the purpose of these gradients is to provide a
means of separating cells according to their densities, the only
necessary property of a gradient for isopycnic centrifugation is
that it contain a range of densities that includes the densities of
the cells to be separated. In the name of brevity, we shall emphasize only one, often overlooked feature of isopycnic gradients. If
there are large differences, with respect to density and viscosity,
between the sample to be layered over the gradient and that part of
the gradient at which the cells enter the gradient, cells will be
compressed and concentrated against this interface just as they are
against the interfaces in discontinuous gradients. While the resultant aggregation of cells has been minimized by some investigators by stirring the interface with a pipette prior to centrifugation, this is not necessary when the gradient is well designed.
     The design of gradients for velocity sedimentation has been
discussed in detail (77). It is worthy of note that the distance of
the gradient from the center of the centrifuge will influence the
degree of separation of cells that is achieved (77,78). As for all
gradients for the separation of cells, the osmolarity of the gradient is important. As discussed by Williams (79), the importance of
diameter in determining the rates of sedimentation of cells has of-

ten been emphasized to an inappropriate degree. Despite this fact,
velocity sedimentation is used most commonly for the separation of
cells that differ with respect to diameter. When this is the goal,
the effect of the diameter of the cell can be maximized and the effect of the density of the cell can be minimized by the design of a
gradient of low density. This has been discussed with some practical examples (28). Resolution will be maximized when the slope
(gm/ml/cm) of the gradient is as small as is consistent with the
stability of the column of fluid (77). When the slope of the gradient is large, rapidly sedimenting cells that have become separated
from less rapidly sedimenting cells are retarded by a barrier of
increasing viscosity and density; this decreases the final distance
between cells. The isokinetic gradient of Ficoll in tissue culture
medium that we described (34,54) combines a low density with a slope
that is sufficiently small as to approach the lower limit of slope
that is compatible with the stability of the gradient. Parentheti-
cally, the low density and low viscosity of this gradient makes it
possible to separate organelles in this gradient by velocity sedimentation with higher resolution and much lower centrifugal forces
than have been used traditionally for the purification of organelles
by velocity sedimentation (77).

Sedimentation Artifacts and Evaluation of Data

The failure of many investigators to be aware of common artifacts in
sedimentation and the lack of a rigorous evaluation of data following sedimentation have resulted in the misinterpretation of a very
large proportion of experiments on the sedimentation of cells. We
have stepped in most of the available potholes at least once and
cannot emphasize too strongly the importance of artifacts that may
occur during sedimentation (3,28). It is very important for anyone
who plans to use velocity sedimentation for the purification of
cells to be aware of the concept of band capacity and the slightly
different concept of gradient capacity.

Relevant to the kinds of data that are required for the evaluation of experiments in cell separation, we (80) wish to emphasize
particular data that are often omitted from published papers that
should be included. Specifically, we believe that it is very important to quantify the markers of function and purity in the separated
fractions and in the starting suspension of cells. For most cell
separations, we would like to see photomicrographs of permanent
preparations of purified cells with sufficient cells being present
to give a semiquantitative message about purity and, more importantly, a clear indication of the quality of the (a) preparations
examined and (b) cells recovered. While morphology of high quality
does not guarantee that the cells are functionally intact or even
alive, it is often possible to tell that cells are damaged or highly
selected based on the morphology of cells in the starting sample as
compared with the quality of the purified cells. Permanent preparations (a) always give a less optimistic estimate of the degree of
purification than wet preparations and (b) provide a permanent
record that can be examined and compared with other experiments in
future years.

Some investigators express contempt for those who are inter-
ested in separating cells of type A from cells of type B while
emphasizing the value of techniques for cell separation to separate
functionally heterogeneous subpopulations of single kinds of cells
such as granulocytes or macrophages. While it is valuable to separate heterogeneous subpopulations of particular kinds of cells that
differ with respect to function, it is of little use to show that
cell of type A is distributed in fractions 6 through 10 and that
these fractions contain different average amounts of function/cell.
If the cells in fraction 6 contain only half as much average func-
tion/cell as cells in fraction 10, does this represent their characteristic heterogeneity in function that was present in vivo, or were
cells in fraction 6 partitioned to fraction 6 because they were injured or possessed an unrelated physical property? How does the
collective function of cells in all fractions compare with the function available before the procedure for purification was performed?
Other similar questions have been discussed in detail (80).

Directions for the Future

It has become increasingly apparent that the optimal purification of
most kinds of cells requires the successive use of more than one
kind of technique for the separation of cells. For example, in purifying the putative precursors of hepatocellular carcinoma from the
livers of rats that had been treated with carcinogen, considerable
advantage was found in the sequential combination of velocity sedimentation and free-flow electrophoresis (81). While electrophoresis
has not been used as extensively as many other approaches to the
separation of cells, electrophoresis is an important approach that
(a) takes advantage of a qualitatively different set of characteristics of cells from those that are important for other cell separations and (b) is generally applicable to all cells (82-84). The development of electrophoretic buffers of low ionic strength and their
application by the late Klaus Zeiller in collaboration with Kurt
Hannig did much to establish the usefulness of electrophoresis for
the separation of large numbers of cells (85-87).
        The number of monoclonal antibodies that are specific for the
surface components of particular kinds of cells is increasing rapidly, and the available technology for the creation of new, highly
specific monoclonal antibodies against specific kinds of cells is
greatly improved (88). These antibodies will be used increasingly
in providing additional, often very specific means for the purification of cells. They may be used to facilitate the electronic sorting of cells (89) or may be used attached to microspheres (90,91)
that will result in changes in charge, density, or susceptibility to
magnetic forces that can be exploited for the purification of the
cell-microsphere complex as described in later chapters. Particles
that can be attracted with magnets have been used for years with
hand-held magnets. More recently, a relatively inexpensive and much
more powerful device (92) has been fabricated that permits one to
separate cells that flow through a column from which they can be
recovered when the electromagnet is switched off.
        We anticipate that sedimentation will continue to provide an
important approach to cell separation that will be combined with

many other qualitatively different approaches to generate strategies for multi-step purifications for the successive exploitation of qualitatively different characteristics of the cells to be separated.

Acknowledgments

This work was supported by Public Health Service grants RO1 CA36467, RO1 CA38727, RO1 CA48032, and P30 CA43703 from the National Cancer Institute and by grant #89B48 from the American Institute for Cancer Research.

Literature Cited

1. Lindahl, P. E. Nature 1948, 161, 648-649.
2. Lindahl, P. E. Biochim. Biophys. Acta 1956, 21, 411-415.
3. Pretlow, T. G.; Weir, E. E.; Zettergren, J. G. Int. Rev. Exp. Pathol. 1975, 14, 91-204.
4. Boone, C. W.; Harell, G. S.; Bond, H. E. J. Cell Biol. 1968, 36, 369-378.
5. Pretlow, T. G.; Boone, C. W. Exp. Mol. Pathol. 1969, 11, 139-152.
6. Pretlow, T. G.; Boone, C. W.; Shrager, R. I.; Weiss, G. H. Anal. Biochem. 1969, 29, 230-237.
7. Leif, R. C.; Vinograd, J. Proc. Natl. Acad. Sci. USA 1964, 51, 520-528.
8. Leif, R. C. In Automated Cell Identification and Cell Sorting; Wied, G. L., Bahr, G. F., Eds.; Academic Press: New York, NY, 1970; pp 21-96.
9. Shortman, K. Annu. Rev. Biophys. Bioeng. 1972, 1, 93-130.
10. Shortman, K. Aust. J. Exp. Biol. Med. Sci. 1968, 46, 375-396.
11. Shimizu, H.; Riley, M. V.; Cole, D. F. Exp. Eye Res. 1967, 6, 141-151.
12. Pertoft, H.; Back, O.; Lindahl-Kiessling, K. Exp. Cell Res. 1968, 50, 355-368.
13. Pertoft, H.; Laurent, T. C. In Cell Separation: Methods and Selected Applications; Pretlow, T. G., Pretlow, T. P., Eds.; Academic Press: New York, NY, 1982, Vol. 1; pp 115-152.
14. Williams, N.; Kraft, N.; Shortman, K. Immunology 1972, 22, 885-899.
15. Abeloff, M. D.; Mangi, R. J.; Pretlow, T. G.; Mardiney, R., Jr. J. Lab. Clin. Med. 1970, 75, 703-710.
16. Pretlow, T. G.; Cassady, I. M. Am. J. Pathol. 1970, 61, 323-339.
17. Pretlow, T. G.; Pichichero, M. E.; Hyams, L. Am. J. Pathol. 1971, 63, 255-276.
18. Pretlow, T. G.; Pushparaj, N. Immunology 1972, 22, 87-91.
19. Pretlow, T. G.; Glick, M. R.; Reddy, W. J. Am. J. Pathol. 1972, 67, 215-226.
20. Stewart, M. J.; Pretlow, T. G.; Hiramoto, R. Am. J. Pathol. 1972, 68, 163-182.
21. Pretlow, T. G.; Luberoff, D. E. Immunology 1973, 24, 85-92.
22. Pretlow, T. G.; Williams, E. E. Anal. Biochem. 1973, 55, 114-122.
23. Blackmon, J.; Pitts, A.; Pretlow, T. G. Am. J. Pathol. 1973, 72, 417-426.
24. Pretlow, T. G.; Scalise, M. M.; Weir, E. E. Am. J. Pathol. 1974, 74, 83-94.

25. Rickwood, D.; Hell, A.; Birnie, G. D.; Gilhuus-Moe, C. C. Biochim. Biophys. Acta **1974**, 342, 367-371.
26. Rickwood, D. Metrizamide, A Gradient Medium for Centrifugation Studies; 3rd Ed.; Nyegaard: Oslo, Norway, 1979.
27. Gleich, G. J.; Ackerman, S. J.; Loegering, D. A. In Cell Separation: Methods and Selected Applications; Pretlow, T. G., Pretlow, T. P., Eds.; Academic Press: New York, NY, 1983, Vol. 2; pp 15-32.
28. Pretlow, T. G.; Pretlow, T. P. In Cell Separation: Methods and Selected Applications; Pretlow, T. G., Pretlow, T. P., Eds.; Academic Press: New York, NY, 1982; Vol. 1; pp 41-60.
29. Pretlow, T. G.; Pretlow, T. P. In Biomembranes; Fleischer, S., Fleischer, B., Eds.; Methods in Enzymology; Academic Press: New York, NY, 1989, Vol. 171; pp 462-482.
30. Vincent, R.; Nadeau, D. Anal. Biochem. **1984**, 141, 322-328.
31. Aspray, D. W.; Pitts, A.; Pretlow, T. G. Anal. Biochem. **1975**, 66, 353-364.
32. Kreisberg, J. I.; Pitts, A. M.; Pretlow, T. G. Am. J. Pathol. **1977**, 86: 591-602.
33. Haskill, J. S.; Moore, M. A. S. Nature **1970**, 226, 853-854.
34. Pretlow, T. G.; Pretlow, T. P. In Cell Separation: Methods and Selected Applications; Pretlow, T. G., Pretlow, T. P., Eds.; Academic Press: New York, NY, 1987, Vol. 5; pp 281-309.
35. Meistrich, M. L. In Cell Separation: Methods and Selected Applications; Pretlow, T. G., Pretlow, T. P., Eds.; Academic Press: New York, NY, 1983, Vol. 2, pp 33-74.
36. Keng, P. C.; Siemann, D. W.; Lord, E. M. In Cell Separation: Methods and Selected Applications; Pretlow, T. G., Pretlow, T. P., Eds.; Academic Press: New York, NY, 1987, Vol. 5; pp 51-74.
37. Wells, J. R.; Opelz, G.; Cline, M. J. J. Immunol. Methods **1977**, 18, 79-93.
38. Daugherty, D. F.; Scott, J. A.; Pretlow, T. G. Clin. Exp. Immunol. **1980**, 42, 370-377.
39. Green, C. L.; Pretlow, T. P.; Tucker, K. A.; Bradley, E. L., Jr.; Cook, W. J.; Pitts, A. M.; Pretlow, T. G. Cancer Res. **1980**, 40, 1791-1796.
40. Miller, S. B.; Pretlow, T. P.; Scott, J. A.; Pretlow, T. G. J. Natl. Cancer Inst. **1982**, 68, 851-857.
41. Pretlow, T. P.; Pretlow, T. G. In Cell Separation: Methods and Selected Applications; Pretlow, T. G., Pretlow, T. P., Eds.; Academic Press: New York, NY, 1983, Vol. 2; pp 221-233.
42. Hymer, W. C.; Hatfield, J. M. In Cell Separation: Methods and Selected Applications; Pretlow, T. G., Pretlow, T. P., Eds.; Academic Press: New York, NY, 1984, Vol. 3; pp 163-194.
43. Bertoncello, I. In Cell Separation: Methods and Selected Applications; Pretlow, T. G., Pretlow, T. P., Eds.; Academic Press: New York, NY, 1987, Vol. 4; pp 89-108.
44. Gillespie, G. J. In Cell Separation: Methods and Selected Applications; Pretlow, T. G., Pretlow, T. P., Eds.; Academic Press: New York, NY, 1982; Vol. 1; pp 61-83.
45. Peterson, E. A.; Evans, W. H. Nature **1967**, 214, 824-825.
46. Miller, R. G.; Phillips, R. A. J. Cell. Physiol. **1969**, 73, 191-202.
47. Mel, H. C. J. Theoret. Biol. **1964**, 6, 159-180.

48. Mel, H. C. J. Theoret. Biol. 1964, 6, 181-200.
49. Mel, H. C. J. Theoret. Biol. 1964, 6, 307-324.
50. Mel, H. C.; Mitchell, L. T.; Thorell, B. Blood 1965, 25, 63-72.
51. Groom, A. C.; Anderson, J. C. J. Cell. Physiol. 1972, 79, 127-138.
52. Bont, W. S.; Hilgers, J. H. M. Prep. Biochem. 1977, 7, 45-60.
53. Wells, J. R. In Cell Separation: Methods and Selected Applications; Pretlow, T. G., Pretlow, T. P., Eds.; Academic Press: New York, NY, 1982, Vol. 1; pp 169-189.
54. Pretlow, T. G. Anal. Biochem. 1971, 41, 248-255.
55. Pretlow, T. G.; Pretlow, T. P. Nature 1988, 333, 97.
56. Sanderson, R. J. In Cell Separation: Methods and Selected Applications; Pretlow, T. G., Pretlow, T. P., Eds.; Academic Press: New York, NY, 1982, Vol. 1; pp 153-168.
57. Pretlow, T. G.; Pretlow, T. P. Cell Biophys. 1979, 1, 195-210.
58. Keng, P. C.; Li, C. K. N.; Wheeler, K. T. Cell Biophys. 1981, 3, 41-56.
59. Keng, P. C.; Wheeler, K. T.; Siemann, D. W.; Lord, E. M. Exp. Cell Res. 1981, 134, 15-22.
60. Keng, P. C.; Siemann, D. W.; Wheeler, K. T. Br. J. Cancer 1984, 50, 519-526.
61. Keng, P. C.; Rubin, P.; Constine, L. S.; Frantz, C.; Nakissa, N.; Gregory, P. Int. J. Radiat. Oncol. Biol. Phys. 1984, 10, 1913-1922.
62. Meistrich, M. L.; Meyn, R. E.; Barlogie, B. Exp. Cell Res. 1977, 105, 169-177.
63. Grabske, R. J.; Lake, S.; Gledhill, B. L.; Meistrich, M. L. J. Cell. Physiol. 1975, 86, 177-190.
64. Wells, J. R.; Opelz, G.; Cline, M. J. J. Immunol. Methods 1977, 18, 63-77.
65. Pretlow, T. G.; Pretlow, T. P. In Patterson, M. K., Jr. Ed.; In Vitro, Monograph #5, Tissue Culture Association: Gaithersburg, MD, 1984, pp 4-18.
66. Pretlow, T. G.; Jones, J.; Dow, S. Am. J. Pathol. 1974, 74, 275-286.
67. Raidt, D. J.; Mishell, R. I.; Dutton, R. W. J. Exp. Med. 1968, 128, 681-698, 1968.
68. Boyum, A. Nature 1964, 204, 793-794.
69. Boyum, A. Scand. J. Clin. Lab. Invest. 1968, 21 (Suppl. 97), 9-29.
70. Boyum, A. Scand. J. Clin. Lab. Invest. 1968, 21 (Suppl. 97), 31-50.
71. Boyum, A. Scand. J. Clin. Lab. Invest. 1968, 21 (Suppl. 97), 51-76.
72. Boyum, A. Scand. J. Clin. Lab. Invest. 1968, 21 (Suppl. 97), 77-89.
73. Boyum, A. Scand. J. Clin. Lab. Invest. 1968, 21 (Suppl. 97), 91-106.
74. Boyum, A. Scand. J. Clin. Lab. Invest. 1968, 21 (Suppl. 97), 107-109.
75. Aiuti, F.; Cerottini, J.-C.; Coombs, R. R. A.; Cooper, M.; Dickler, H. B.; Froland, S. S.; Fudenberg, H. H.; Greaves, M. F.; Grey, H. M.; Kunkel, H. G.; Natvig, J. B.; Preud'homme, J.-L.; Rabellino, E.; Ritts, R. E.; Rowe, D. S.; Seligmann, M.;

    Siegal, F. P.; Stjernsward, J.; Terry, W. D.; Wybran, J. Scand.
    J. Immunol. 1974, 3, 521-532.
76. de Duve C. J. Cell Biol. 1971, 50, 20d-55d.
77. Pretlow, T. G.; Kreisberg, J. I.; Fine, W. D.; Zieman, G. A.;
    Brattain, M. G.; and Pretlow, T. P. Biochem. J. 1978, 174,
    303-307.
78. Pretlow, T. G.; Boone, C. W. Science 1968, 161, 911-913.
79. Williams, N. In Cell Separation: Methods and Selected Appli-
    cations; Pretlow, T. G., Pretlow, T. P., Eds.; Academic Press:
    New York, NY, 1982, Vol. 1; pp 85-113.
80. Pretlow, T. G.; Pretlow, T. P. In Cell Separation: Methods and
    Selected Applications; Pretlow, T. G., Pretlow, T. P., Eds.;
    Academic Press: New York, NY, 1982, Vol. 1; pp 31-40.
81. Miller, S. B.; Saccomani, G.; Pretlow, T. P.; Kimball, P. M.;
    Scott, J. A.; Sachs, G.; Pretlow, T. G. Cancer Res. 1983, 43,
    4176-4179.
82. Hannig, K. In Electrophoresis. Theory, Methods, and Applica-
    tions; Bier, M., Ed.; Academic Press: New York, NY, 1967, Vol.
    2; pp 423-471.
83. Pretlow, T. G.; Pretlow, T. P. Intl. Rev. Cytol. 1979, 61,
    85-128.
84. Platsoucas, C. D. In Cell Separation: Methods and Selected
    Applications; Pretlow, T. G., Pretlow, T. P., Eds.; Academic
    Press: New York, NY, 1983, Vol. 2; pp 145-182.
85. Zeiller, K.; Schubert, J. C. F.; Walther, F.; Hannig, K.
    Hoppe-Seyler's Z. Physiol. Chem. 1972, 353, 95-104.
86. Zeiller, K.; Loser, R.; Pascher, G.; Hannig, K. Hoppe-Seyler's
    Z. Physiol. Chem. 1975, 356, 1225-1244.
87. Zeiller, K.; Hansen, E. J. Histochem. Cytochem. 1978, 26,
    369-381.
88. Barald, K. F. In Cell Separation: Methods and Selected Appli-
    cations; Pretlow, T. G., Pretlow, T. P., Eds.; Academic Press:
    New York, NY, 1987, Vol. 5; pp 89-102.
89. Preffer, F. I.; Colvin, R. B. In Cell Separation: Methods and
    Selected Applications; Pretlow, T. G., Pretlow, T. P., Eds.;
    Academic Press: New York, NY, 1987, Vol. 5; pp 311-347.
90. Molday, R. S. In Cell Separation: Methods and Selected Appli-
    cations; Pretlow, T. G., Pretlow, T. P., Eds.; Academic Press:
    New York, NY, 1984, Vol. 3; pp 237-263.
91. Owen, C. S.; Liberti, P. A. In Cell Separation: Methods and
    Selected Applications; Pretlow, T. G., Pretlow, T. P., Eds.;
    Academic Press: New York, NY, 1987, Vol. 4; pp 259-275.
92. Owen, C. S. In Cell Separation: Methods and Selected Appli-
    cations; Pretlow, T. G., Pretlow, T. P., Eds.; Academic Press:
    New York, NY, 1983, Vol. 2; pp 127-144.

RECEIVED March 15, 1991

Chapter 7

High-Capacity Separation of Homogeneous Cell Subpopulations by Centrifugal Elutriation

Peter C. Keng

Department of Radiation, Oncology, and Experimental Therapeutics, University of Rochester School of Medicine and Dentistry, Rochester, NY 14642

The feasibility of obtaining a very large quantity ($>10^8$) of homogeneous cell populations by the Beckman JE-10X elutriator system was evaluated by 5-15 μm latex spheres and Chinese hamster ovary (CHO) cells. The evaluation was based on 1) recovery, 2) elution loss during loading, 3) homogeneity of the size distributions, and 4) the relationship of the median volume of eluted particles or cells to the rotor speed and the collection fluid velocity. The results indicated that a minimum number of 2×10^8 cells was required for any reasonable separation. More than 90% of the particles or cells were always recovered when more than 5×10^8 total cells were loaded into the system. The homogeneity of separated fractions as determined by the coefficient of variation (CV) of volume distributions showed a limiting CV of 7% for latex spheres and 11.5% for CHO cells. Fractions containing 95% G_1 cells, 82% S cells, and 78% G_2M cells were obtained from asynchronous CHO cells with this method.

There are many theoretical questions and practical problems in cell biology that require studies of large numbers of relatively homogeneous cell subpopulations. Centrifugal elutriation has been available as a cell separation technique since 1948 when Lindahl first reported the principle of "counter-streaming centrifugation". Lindahl and his co-workers used this technique to separate cell subpopulations from a variety of biological systems (1-4). However, it was not until 1973, when the Beckman Co. introduced the JE-6 elutriator rotor, that the centrifugal elutriation technique became generally available.

The principle of cell separation by centrifugal elutriation has been described in detail elsewhere (1, 5-8). Briefly, cells suspended in the separation chamber of an elutriator rotor are subjected to two opposite forces: the centrifugal force generated from the rotation of the rotor in an outward direction and the fluid force through the chamber in an inward or centripetal direction. The special shape of the separation chamber makes it possible to generate a gradient of flow velocity, decreasing toward the center of rotation. Cells of a

0097–6156/91/0464–0103$06.00/0

given size are sedimented to a position in the separation chamber according to their sedimentation velocity, which is based on the size, density, and shape of the cell. The cells remain in the chamber as long as the two opposite forces are in balance. By incremental increase in the flow rate of the fluid or by decrease in the centrifugal force, distinct populations of cells with relatively homogeneous sizes can be eluted out of the rotor sequentially.

Centrifugal elutriation has been used successfully in many cell separation studies to provide adequate numbers (10^5-10^6 cells) of homogeneous cells (7, 8). Recently, there is an increased demand for a large quantity ($>10^8$ cells) of homogeneous ($>90\%$ purity) cells in many biological and biomedical studies. For example, purification of specific proteins that regulate the progression of cells into different phases of the cell cycle requires 10^7 to 10^9 synchronized G_1, S and G_2M cells (9, 10). The isolation of normal bone marrow stem cells used for autologous marrow transplantation in cancer patients requires the ability to separate 10^9 to 10^{10} cells in a relatively short period of time (11). To fulfill this requirement, Beckman has recently developed a large capacity elutriator rotor with a separation chamber ten times the size of that used in the JE-6 system. However, little information is available about the theoretical and practical aspects of this new elutriator system. In this study, we have used latex spheres and Chinese hamster ovary (CHO) cells as test models to evaluate the cell loading capacity, coefficient of variation (CV) of volume distributions, and overlapping of volume distributions from separated fractions. Finally, the degree of synchrony in isolated G_1, S, and G_2M CHO cells was determined by flow cytometric analysis.

Materials and Methods

Cells and Latex Spheres. Chinese hamster ovary (CHO) cells were maintained in culture with F-10 medium supplemented with 10% (v/v) fetal bovine serum (Grand Island Biological Co., Grand Island, NY) as described previously (10). Exponentially growing cells were trypsinized from flasks and suspended in ice-cold F-10 medium with serum before elutriation.

Polystyrene divinyl benzene (latex) spheres of uniform shape and density (1.05 g/cm^3) with diameters of individual samples ranging from 5 to 15 μm were purchased from Sigma Chemical Co. (St. Louis, MO) for elutriation experiments. Latex spheres were suspended in 0.9% NaCl solution at a concentration of 1 x 10^8 particles/mL and stored at room temperature as stock solution.

Centrifugal Elutriation. Single cell suspensions of CHO cells and latex particles were separated using the Beckman JE-10X rotor driven by a Beckman J6 M centrifuge. The centrifuge speed was controlled through the digital speed-control board (Beckman Instrument Inc., Palo Alto, CA) to allow the rotor speed selection to within ± 10 revolutions per minute (RPM). Rotor speed was measured directly by the electronic detector connected to the driver motor. Fluid flow through the elutriator system was maintained by a Multiperpex pump with fine velocity control (LKB Instruments, Baltimore, MD). The flow rate was continuously monitored using an in-line flow meter (VWR Scientific Co., Rochester NY).

Prior to elutriation, the rotor was checked for possible external or internal leakage through defective seals or O-rings as described previously (*8*). No experiments were performed until all leakage problems were corrected. Cells were elutriated in ice-cold complete F-10 medium with serum, whereas latex spheres were elutriated at room temperature in 0.9% sodium chloride in water. In the initial studies, the fluid was pumped through the elutriator system at 50 mL/min with the rotor speed set at 2000 ± 10 rpm for both cells and latex spheres. The initial conditions were chosen such that the resulting centrifugal and fluid forces would cause cells or latex spheres with their respective sizes and densities to be retained in the separation chamber (*1, 7*). During and immediately after the loading, two 50-mL fractions of eluted fluid were collected. Thereafter, four 50-mL fractions were collected at each rotor speed. The rotor speeds were selected to allow a constant increment in elutriated cell volume (1.5 x) for each consecutive collection (see elutriation equations below). In addition to the 50 mL/min flow rate, CHO cells were also separated with a variable flow rate to determine the effect of fluid flow rate on the separation process. When different flow rates were used, the rotor speed was kept at a constant value of 850 rpm (152 xg).

Cell and Particle Analysis. Each fraction collected from the elutriator, as well as a portion of the initial unfractionated sample, was analyzed for the number of cells or particles and for their size distribution using a Coulter Counter Channelyzer system (Coulter Electronics, Hialeah, FL). The system was equipped with a 100 μm orifice tube and operated with the edit circuit switch on. The median volume of the cell or particle population was determined from the median channel number of the volume distribution using a calibration constant derived from latex spheres of uniform size and density.

Elutriation Equation. The physical theory of centrifugal elutriation was described initially by Lindahl (*1*) and later by others (*5, 6*). In order to evaluate the separation characteristics of the Beckman JE-10X elutriator rotor, the relationship between the elutriated cell volume (V) and the collection rotor angular velocity (ω) or the fluid flow velocity (v) must be established. This can be done by equilibrating the three forces; F_c (centrifugal force), F_b (buoyant force), and F_d (fluid drag force); acting on the spherical particle with diameter d (*5*).

$$F_c = (\pi/6)d^3 \rho \omega^2 R \tag{1}$$

$$F_b = -(\pi/6)d^3 \rho_o \omega^2 R \tag{2}$$

$$F_d = -3\pi\eta dv \tag{3}$$

where ω is the rotor angular velocity, R is the distance from the particle to the center of the rotor, ρ_o is the fluid density, ρ is the particle density, η is the fluid viscosity, and v is the local fluid velocity.

When a particle or cell is at steady state in the chamber:

$$F_c + F_b + F_d = 0 \tag{4}$$

and the volume of the particle being removed from the separation chamber is:

$$V = k\omega^{-3}$$

where

$$k = 9\pi\sqrt{2}[\upsilon\eta/R(\rho-\rho_o)]^{3/2} \tag{5}$$

or

$$V = k'\upsilon^{1.5}$$

where

$$k' = 9\pi\sqrt{2}[\eta/R(\rho-\rho_o)]^{3/2}\omega^{-3} \tag{6}$$

If all of the above parameters but one are known, the remaining parameter can be calculated.

Flow Cytometry.   Flow cytometry was used in an attempt to monitor the percentage of separated CHO cells in the G_1, S, or G_2M phases of the cell cycle. For flow cytometric analysis, cells were first removed from the medium by gentle centrifugation, fixed in 70% ethanol and then treated with RNAse (1 mg/mL, 30 min) and stained with propidium iodide (PI, 10 μg/ml) as previously described (9).   Cellular DNA content was measured using a Coulter EPICS Profile flow cytometer with 25 mW of power at an excitation wavelength of 488 nm.   The fraction of cells in the G_1, S, and G_2M phases of the cell cycle was determined by computer analysis of the DNA histograms using mathematical models described by Dean and Jett (12) and Fried (13).

Results

Recovery from the Elutriator.   Usually >90% of the CHO cells or latex spheres loaded into the separation chamber were recovered in each elutriation experiment when more than 2×10^8 cells or particles were loaded; always, this was the case when more than 5×10^8 cells or particles were loaded (Figure 1, 2). The first two 50-mL fractions collected during and immediately after loading contained about 5% of all the cells or particles loaded.   These cells or particles had a volume distribution similar to that of the unseparated sample.   However, if less than 2×10^8 cells or particles were loaded, the recovery varied from 35 to 60% and there was no separation of different sized particles in the collected fractions (Figure 1).   Usually the first two fractions contained 30 to 50% of total loaded cells and each remaining fraction contained 5×10^3 to 1.5×10^4 cells with a volume distribution similar to that of the unseparated population (Figure 1).   By loading various numbers of cells or particles into the JE-10X rotor, we have determined that the optimum number of cells for an ideal separation was between 5×10^8 and 2×10^9 cells (Figure 2, 3).

Conformation to the Elutriation Equation.   Since non-ideal recovery phenomena were observed, it was necessary to investigate whether the ideal separation could

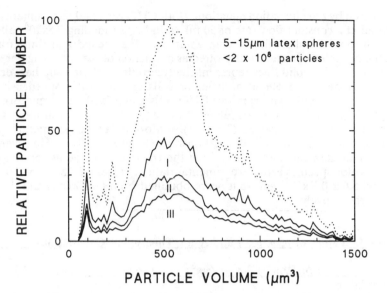

Figure 1. Volume distributions of separated and unseparated 5-15 μm diameter latex spheres. 1 x 10^8 spheres were loaded into the elutriator system. Dotted curve is for unseparated latex spheres and solid curves I, II, and III correspond to fraction I, II, and III in Table II, respectively.

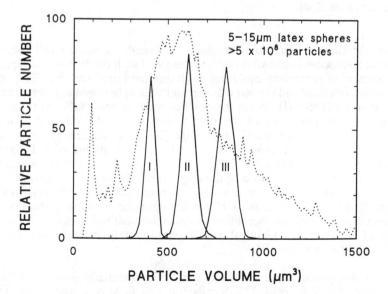

Figure 2. Volume distributions of separated and unseparated 5-15 μm diameter latex spheres. 8.5 x 10^8 spheres were loaded into the elutriator system. Dotted and solid curves represent unseparated spheres and fraction I, II, and III respectively.

be predicted by the elutriation equation. Both CHO cells and latex spheres were separated at a constant flow rate of 50 mL/min. After loading 7.5×10^8 cells into the separation chamber, the median volume of the second and third fraction collected at each rotor speed was plotted as a function of the rotor angular speed (Figure 4, 5). The solid lines represent the theoretical relationship between the median volume of the eluted particles or cells and the rotor speed using the constants, $k = 8.7 \times 10^{11} \ \mu m^3 \ rpm^3$ for CHO cells and $5.5 \times 10^{11} \ \mu m^3 \ rpm^3$ for latex spheres. These values were calculated from equation 5 by substituting known values of η, ρ, ρ_o, v, and R (Table I). Most of the experimental points conformed to the theoretical relationship (Figure 4, 5). However, the experimental data deviated slightly from the theoretical curve at very high and very low rotor speeds. From experimental data, k values of $8.5 \pm 0.5 \times 10^{11} \ \mu m^3 \ rpm^3$ and $5.6 \pm 0.3 \times 10^{11} \ \mu m^3 \ rpm^3$ were obtained from CHO cells and the latex spheres, respectively.

Table I. Values of v, ρ, ρ_o, η and R for Theoretical Calculations

Cell type	v (mL/min)	ρ_o (g/cm^3)	ρ (g/cm^3)	η (cP)	R (cm)
5-15 μm latex spheres[a]	50	1.00365	1.053	1.0019	18.8
CHO cells[b]	50	1.00870	1.075[c]	1.870	18.8

a In 0.9% NaCl at 20°C.
b In complete F-10 with serum at 4°C.
c Literature cited #8

Size Distributions of Separated Fractions. Although the median cell volume of separated latex spheres and CHO cells agreed well with the theoretical equation, the resolution of separation could not be determined until the size distributions of separated fractions and the cell volume overlapping between adjacent fractions were measured (Table II). When the coefficient of variation (CV) of the volume distribution from latex spheres was determined, a constant value of about 7% was obtained in various fractions (Figure 2). The degree of cell volume overlapping between successive fractions with relative volume of 1, 1.5 and 2.0 was between 3 to 5%. For CHO cells, the CV (11.5%) and volume overlapping (5-20%) were higher than those of latex spheres. The difference between latex spheres and cells is probably due to the inherent biological variations in shape and density among CHO cells. CHO cells separated by varying the flow rate were also tested to see whether different collection methods would influence the quality of separation. Identical results were obtained from two different collection methods (Table II).

Cell Cycle Analysis of Separated CHO Cells. Exponentially growing CHO cells that contained 52% G_1 cells, 32% S cells and 16% G_2M cells were separated by centrifugal elutriation using a constant flow rate of 50 mL/min (Table II). The DNA histograms were obtained from fraction I, II, and III with a median cell

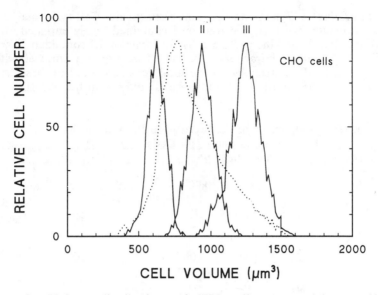

Figure 3.   Volume distributions of CHO cells separated by centrifugal elutriation using a constant flow rate of 50 mL/min and different rotor speeds as described in Table II. 1 x 10⁹ cells were loaded into the elutriator system. Dotted curve is obtained from unseparated exponentially growing cells. Solid curves correspond to fraction I, II, and III in Table II.

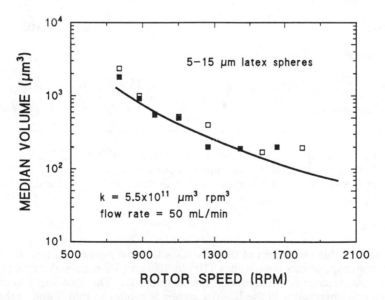

Figure 4.   Median volume of 5-15 μm latex spheres as a function of the collection rotor speed after elutriation with a flow rate of 50 mL/min. Open and closed squares are second and third fractions respectively. Solid curve is calculated by equation 5 with parameters listed in Table I.

volume of 625 μm^3, 937 μm^3, and 1250 μm^3 respectively. Cell cycle analysis with the flow cytometer indicated that fraction I contained highly enriched G_1 cells (>95%); fraction II contained 82% S cells; and fraction III contained 78% G_2M cells (Figure 6). Although fractions I, II, and III were greatly enriched with G_1, S, and G_2M cells, the percentage of the total cells recovered in these fraction was from 20% in fraction I, 12% in fraction II, and 6% in fraction III.

Table II. Comparison of the Separation Parameters for Latex Spheres and CHO Cells

Collect Method	5-15 μm Latex Spheres			CHO Cells		
Rotor speed[a] (RPM)	1110	975	890	1110	975	890
Fraction	I	II	III	I	II	III
Median volume (μm^3)	410	605	800	625	937	1250
CV, %	7.1	7.2	7.1	11.5	11.8	11.7
Volume overlap[b] %	2.1	5.1	3.6	5.1	20.5	12.3
Flow rate (mL/min)[c]				30	38	46
Fraction				I	II	III
Median volume (μm^3)				638	927	1235
CV, %				11.3	12.1	11.6
Volume overlap %				5.5	22.1	13.2

a Constant flow rate of 50 mL/min, total collection of 200 mL fluid.
b % cell volume extended to adjacent fractions (I, II, III).
c Constant rotor speed of 850 rpm, total collection of 200 mL fluid.

Discussion

The results of this evaluation of the JE-10X elutriator system indicate that very large homogeneous cell populations (10^7 to 10^8) can be obtained when proper loading and separation procedures are employed. The non-ideal recovery phenomenon observed with the JE-10X system is similar to that found in the JE-6 system. The possible causes of non-ideal recovery phenomena are fluid flow disturbances in the separation chamber, which at the present time, we can not

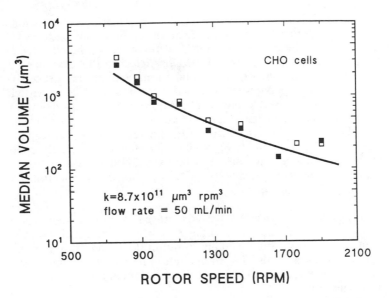

Figure 5. Median volume of CHO cells as a function of the collection rotor speed after elutriation with a flow rate of 50 mL/min. Open and closed squares are second and third fractions, respectively. Solid curve is calculated by equation 5 with parameters listed in Table I.

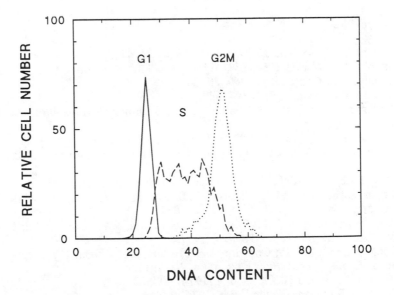

Figure 6. DNA histograms of separated CHO cells obtained from Figure 3. G_1, S, and G_2M fractions correspond to fraction I, II, and III respectively.

confirm in the JE-10X separation chamber. The orientation and location of the chamber in the rotor make it difficult to analyze the fluid flow patterns with dye solutions. However, the problem of non-ideal recovery can be circumvented by using the appropriate number of cells or particles in the chamber. In the experiments with latex spheres, substantial improvements in the cell volume distributions and overlapping were obtained when the number of particles was increased from $2x10^8$ to $5x10^8$ (Fig. 1, 2). Another factor that determines the resolution of separation is the volume increment of the cells removed. To collect equally distributed volume increments (i.e., each sample corresponding to an increase of 200 μm^3 in volume), smaller increments in the dropped rotor speed or in the increased flow rate must be used. Using these approaches, we have obtained a good yield (3.75 x 10^7 to 1.5 x 10^8 cells) of synchronized CHO cells (Fig. 6).

Centrifugal elutriation offers the advantages over some methods for cell separation that cells are suspended in any suitable medium such as an isotonic buffer solution or tissue culture medium to retain their viability and to minimize cell loss. Consequently, the functional and ultrastructural integrity of the elutriated cells is preserved, and highly purified cell preparations with satisfactory recovery can be obtained.

Acknowledgments

This work was supported in part by grants from the National Cancer Institute of the National Institutes of Health, CA-11198 and CA-44723, and CA-11051. We thank B. King for technical support.

Literature Cited

1. Lindahl, P.E. *Biochem. Biophys. Acta* **1956,** *21,* 411.
2. Lindahl, P.E. *Nature* **1956,** *178,* 491.
3. Lindahl, P.E. *Cancer Res.* **1960,** *20,* 841.
4. Lindahl, P.E. *Nature* **1962,** *194,* 589.
5. Sanderson, R.J.; Bird, K.E.; Palmer, N.F.; Brenman, J. *Anal. Biochem.* **1976,** *71,* 615.
6. Pretlow, T.G.,II; Pretlow, T.P. *Cell Biophys.* **1979,** *1,* 195.
7. Meistrich, M.L.; Grdina, K.J.; Meyn, R.E.; Barlogie, B. *Cancer Res.* **1977,** *37,* 4291.
8. Keng, P.C.; Li, C.K.N.; Wheeler, K.T. *Cell Biophys.* **1981,** *3,* 41.
9. Brown, E.H.; Iqbal, M.A.; Stuart, S.; Hatton, K.S.; Valinsky, J.; Schildkraut, C.L. *Mol. Cell Biol.* **1987,** *7,* 450.
10. Liu, Y.; Marraccino, R.L.; Keng, P.C.; Bambara, R.A.; Lord, E.M.; Chou, W.; Zain, S.B. *Biochem.* **1989,** *28,* 2967.
11. Keng, P.C.; Rubin, P.; Constine, L.S.; Frantz, C.; Nikissa, N.; Gregory, P. *Int. J. Radiat. Oncol. Biol. Phys.* **1984,** *10,* 1913.
12. Dean, P.N.; Jett, J.H. *J. Cell Biol.* **1974,** *60,* 523.
13. Fried, J. *Computers Biomed. Res.* **1976,** *9,* 263.

RECEIVED March 15, 1991

Chapter 8

Cell Separations Using Differential Sedimentation in Inclined Settlers

Robert H. Davis, Ching-Yuan Lee, Brian C. Batt, and
Dhinakar S. Kompala

Department of Chemical Engineering, University of Colorado,
Boulder, CO 80309-0424

We are using settling channels with walls inclined from the vertical to separate cell subpopulations on the basis of differences in cell size, density, or aggregation properties. Inclined settlers are particularly suitable for such separations because of their high throughput, simple construction, and ability to be operated on a continuous basis. A steady-state theory to describe cell separations using a continuous inclined settler has been developed, and this theory has been verified for the separation of two yeast strains having slightly different cell sizes. Further experiments have been performed which demonstrate that inclined settlers are capable of separating nonviable hybridoma cells from viable hybridoma cells. We have also shown the potential application of this technology in biotechnology by using it for selective cell separations and recycle during continuous bacterial, yeast, and hybridoma fermentations.

Separating cells from media is important both as a first step in product recovery and in "feed-and-bleed" bioreactors with high cell concentrations. *Selective* cell separations, in which different subpopulations of cells are separated from one another, are even more important to modern biotechnology because they can be used to remove unproductive or parasitic cells from bioreactors or cell culture systems. There is a critical need for selective cell separation methods that can be employed economically on the large-scale. For example, a relatively small, 10-liter fermentor operated at a dilution rate of 0.1 hr^{-1} and a cell concentration of 10^9 cells/ml would require a cell sorter with the ability to process 10^{12} cells/hr in order to selectively separate productive cells from nonproductive cells.

Because of the importance to hematology and other fields, a host of selective cell separation methods has been developed, including differential sedimentation (*1*) and centrifugation (*2*), partitioning in two-phase aqueous systems (*3,4*),

0097–6156/91/0464–0113$06.00/0

affinity adhesion (5,6), magnetic cell sorting (7,8), electrophoresis (9,10), field flow fractionation (11), and optical cell–sorting (12,13). Most of these methods are reviewed elsewhere in this book. In general, they have been developed for analytical purposes and have been limited to small–scale applications (14). However, recent research efforts have included the investigation of high–capacity separations of cell subpopulations because of their importance in biotechnology and other fields.

At the University of Colorado, we are using settling channels with walls inclined from the vertical to separate cells on the basis of differences in sedimentation velocities. Inclined settlers are particularly suitable for high–capacity separations of cell subpopulations because of their simple construction, ease of scaleup, and ability to be operated on a continuous basis. Their capacity or throughput is limited, however, by the small sedimentation velocities of single cells under normal gravity. For example, the sedimentation velocity of a typical brewer's yeast cell in an aqueous medium is approximately 1 cm/hr (15). We have overcome this problem by working with flocculent cells which form aggregates that settle one or more orders–of–magnitude faster than single cells. By genetically coupling the flocculation character of the cells with their ability to overproduce a valuable biological, an inclined settler may be used as a powerful tool which selectively separates the productive, flocculent cells from the unproductive, nonflocculent cells in the product stream of a bioreactor. The productive cells are then recycled to the bioreactor, whereas the unproductive cells are discarded. This strategy has been successfully demonstrated for nonsegregating yeast cultures (16) and for segregating bacterial cultures (17). In the segregating bacterial cultures, plasmid instability causes a continuous culture to be taken over by unproductive, plasmid–free cells after several generations when no selective separation and recycle is used, whereas the culture may be maintained as highly productive for prolonged periods when the inclined settler is employed (18).

In this chapter, we present more recent research in which an inclined settler is used to separate nonflocculent cells on the basis of size or density differences. Two model systems are used, one consisting of nondividing and dividing yeast cells and the other of viable and nonviable hybridoma cells. We first review the theory of inclined settling for selective cell separations. The materials and methods are then described, followed by the results and discussion. We conclude with some remarks about the practical implications of this selective cell separation technology.

Inclined Settler Theory

The theory of an inclined settler operated under steady–state and transient conditions for particle classification is discussed in detail by Davis et al. (19). Here we consider steady–state operation of an inclined settler for selective cell separations, as shown in Figure 1. A cell suspension is fed into the settler at the volumetric flowrate Q_f. This feed suspension contains cells with a distribution

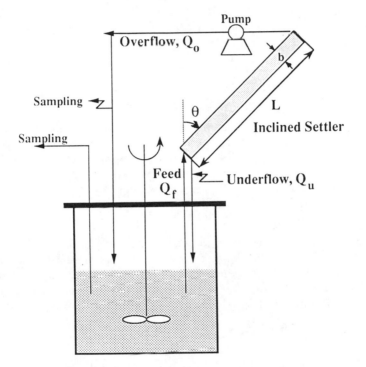

Figure 1.   Schematic of an inclined settler used for selective cell separations.

of settling velocities owing to their distribution of sizes, densities, and/or shapes. This distribution is then separated by the inclined settler into an overflow stream containing a fine fraction of slower–settling cells which did not settle out of suspension in the inclined settler, and an underflow stream containing a coarse fraction of faster–settling cells which did settle out of suspension. The overflow rate and underflow rate are denoted as Q_o and Q_u. Our goal is to predict the composition of the fine and coarse fraction, given the composition of the feed suspension. The equations derived below are for a single subpopulation of cells. They may be applied to each subpopulation present in the feed suspension, or to the entire cell population.

Steady–state mass balances on total suspension, total cells in the subpopulation, and those cells in the subpopulation which have settling velocity v are given as follows:

$$Q_f = Q_o + Q_u \tag{1}$$
$$Q_f X_f = Q_o X_o + Q_u X_u \tag{2}$$
$$Q_f X_f P_f(v) = Q_o X_o P_o(v) + Q_u X_u P_u(v) \tag{3}$$

where X is the total cell mass concentration of the subpopulation and the subscripts f, o, and u refer to the feed, overflow, and underflow streams, respectively. Also, $P(v)$ is the normalized probability density function, defined such that $P(v)dv$ is the fraction of cells by mass in a given stream that have settling velocities between v and $v + dv$. $P(v)$ is normalized so that

$$\int_0^\infty P(v)dv = 1 \tag{4}$$

The probability density function in the feed stream, $P_f(v)$, can be determined either directly by allowing a sample to sediment in the presence of an optical density recording device (15), or indirectly by measuring the particle size distribution with a particle size analyzer and then using an expression such as Stokes' law to relate the sedimentation velocity of each particle to its size:

$$v = \frac{k\, d^2(\rho_c - \rho)\, g}{18\, \eta} \tag{5}$$

where d is the equivalent cell diameter, ρ_c is the cell density, ρ is the fluid density, η is the fluid viscosity, and g is the gravitational acceleration constant. The correction factor k takes into account wall effects, hindered settling effects, irregularity of cell shapes, and any calibration differences in the measured diameter and the appropriate effective diameter for the Stokes' settling velocity.

The rate of sedimentation in an inclined channel can be described by the so–called PNK theory (20). This theory states that the volumetric rate of production of fluid that is clarified of cells with settling velocity v is equal to the vertical settling velocity of the cells multiplied by the horizontal projected area of the upward–facing surfaces of the channel onto which the cells may settle. By referring to Figure 1, this is given as

$$S(v) = v\, w(L\, \sin\theta + b\, \cos\theta) \tag{6}$$

—acid peptide.
ids. When cell
ecome irregular
ut the plasmid

plates from the vertical, and b, w, and
e rectangular settler, respectively.
t on cells entering the overflow stream.
d of a mixture of clarified fluid, which
$S(v)$, and unsettled suspension, which
$Q_o - S(v)$. A cell mass balance on the

e 180XY parti-
erage equivalent
for nondividing
factor in Stokes
iding cells (*19*),
account for their

$$Q_o > S(v) \qquad (7)$$

bension before reaching the overflow,

.2, derived from
enzene–arsonate,
ter histograms of
how considerable
able cells and the
section). As mea-
e diameter of the
meter of the viable

umed that the cells settle without
ation may be applied to any given
ation that the total cell concentra-
s having settling velocities between

$$(v) \qquad S(v) < Q_o \qquad (8)$$
$$S(v) \geq Q_o \qquad (9)$$

ined settlers were
.5 cm, width $w =$
ons), and length L
ations). The angle
rations) or $\theta = 30°$

velocities in the subpopulation,
tion 4 applied, an expression for
n is obtained as

$$P_f(v)\, dv \qquad (10)$$

ormed using a total
e aqueous medium
ensity and viscosity
rst grown batchwise
concentration, the
the suspension was
. Cells that settled
ward–facing wall of
ity and were thereby
led to the bioreactor
systematically. After
recycle system was
er to assure steady-

defined by $S(v_o) = Q_o$. Thus,
ating $S(v_o)$ with the overflow
Cells with settling velocities
re reaching the overflow. The
erflow stream is obtained by

$$v < v_o \qquad (11)$$

$$v \geq v_o \qquad (12)$$

re performed in a sim-
which fresh nutrient

ow stream are obtained by

cerevisiae 378 with the
mutation which enables

cell divisions to be stopped in the presence of α–factor, a 13-amin
This strain cannot grow in a Ura$^-$ medium if cells lose the plasm
divisions are stopped, the cells begin to increase their sizes and b
in shape. The dividing yeast strain is the same host but with
and without the addition of α–factor during their growth phase.

Size distributions of the yeast cells were measured by an Elzo
cle size analyzer manufactured by Particle Data, Inc. Typical av
sphere diameters are 4.5 μm for normal dividing cells and 7.4 μm
cells. The cell density is $\rho_c = 1.13$ g/cm^3 (21). The correction
law was determined to be $k = 0.7$ for the nearly spherical di
whereas $k = 0.5$ was chosen for the nondividing cells in order to
irregular shape.

Hybridoma Cells. A mouse hybridoma cell line, AB2–14
Sp2/0 myeloma and which produces IgG2a antibodies against b
was used. Correlated forward–angle and right–angle light scat
the cells were made using an EPICS flow cytometer. These s
resolution between the size distributions of the smaller nonvi
larger viable cells (see Figure 6 in the Results and Discussion
sured by the Elzone 180XY particle size analyzer, the averag
nonviable cells is approximately 8 μm, whereas the average dia
cells is approximately 13 μm.

Inclined Settler Dimensions and Operation. The inc
fabricated from rectangular glass tubing having height $b = $
4.0 cm (yeast separations) or $w = 5.0$ cm (hybridoma separat
$= 20$ cm (yeast separations) or $L = 23$ cm (hybridoma separ
of inclination from the vertical was set at $\theta = 45°$ (yeast sepa
(hybridoma separations).

Selective separations experiments for yeast cells were perf
cell recycle configuration, as shown in Figure 1. The dilu
employed was maintained at a temperature of 30°C, and its
were nearly equal to those of pure water. The cells were fi
in a 1–2 liter stirred bioreactor. After reaching the desire
culture medium was replaced with a nongrowth buffer, and
drawn through the inclined settler using a peristaltic pum
out of suspension formed a thin sediment layer on the u
the settler. These cells then slid down the wall due to grav
returned to the bioreactor. The overflow stream was recyc
through a separate port. The overflow rate, Q_o, was varied
each adjustment in the overflow rate was made, the tot
allowed to operate for at least 3 hr before sampling in or
state conditions.

The selective separation experiments for hybridomas we
ilar fashion, except that a chemostat operation was used i

medium was continuously added to the bioreactor while a product stream was removed at an equal rate. This was necessary in order to maintain the viable cell population at a nearly constant cell size distribution over the duration of the experiments. Also, the temperature in the reactor and settler was controlled at 37°C.

After steady state was reached for each overflow rate, samples were taken from both the bioreactor and the overflow stream. Since the bioreactor was well-stirred, samples from it are representative of the feed stream. The samples were analyzed by the Elzone 180XY for cell concentrations and size distributions. The viability fraction of the hybridomas was determined by trypan blue staining. The fraction of nondividing yeast cells was determined by replica plating on first a rich medium and then a defined Ura⁻ medium.

Results and Discussion

Selective Separation of Yeast Cells. The size distributions of yeast cells in the feed stream and the overflow stream for a typical experiment are shown in Figure 2. The feed stream consisted of 50% dividing cells (by number) and 50% nondividing cells. The partially resolved peaks for these two subpopulations are apparent in the feed distribution shown as the solid line in Figure 2. This suspension was drawn through the settler at the rate $Q_o = 0.62$ ml/min, with the settler angle from the vertical being $\theta = 45°$. These conditions correspond to a cutoff diameter of $d_o = 6.2$ μm for the largest cells predicted to reach the overflow without settling. Thus, the overflow should contain primarily the smaller, dividing cells. Indeed, the overflow was found to contain 89% dividing cells for this particular experiment. The measured size distribution in the overflow (dashed line in Figure 2) is in very good agreement with the predictions of equation 11 (dotted line in Figure 2). However, a small concentration of cells larger than the theoretical cutoff size did reach the overflow. This may be due to a hydrodynamic dispersion phenomenon that occurs during sedimentation (22) or to the fact that any elongated, nondividing cell with its major axis oriented normal to the gravity vector will settle slower than predicted by equation 5.

Figures 3 and 4 show how the separation of a 50:50 mixture (by number) of dividing and nondividing yeast cells depends upon the overflow rate from the settler. Figure 3 is a plot of the total cell mass concentration (dividing and nondividing cells combined) in the overflow, normalized by the feed concentration, versus the overflow rate from the settler. At low overflow rates, the cell concentration in the overflow was very small because there was sufficient holdup time in the settler for most of the cells to settle out of suspension. As the overflow rate was increased, the cell concentration in the overflow increased because there was less holdup time for settling. The experimental data follow closely the prediction of equation 10, except that a slightly greater concentration of cells reached the overflow for the reasons discussed previously.

Of more direct relevance for selective cell separations is Figure 4, which shows the fraction of nondividing cells in the overflow stream versus the overflow rate.

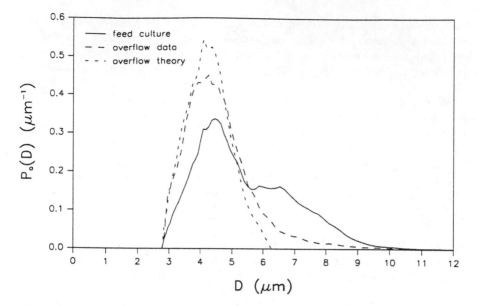

Figure 2.   Size distributions of yeast cells in the feed stream and settler overflow streams for a mixed culture of dividing and nondividing cells.

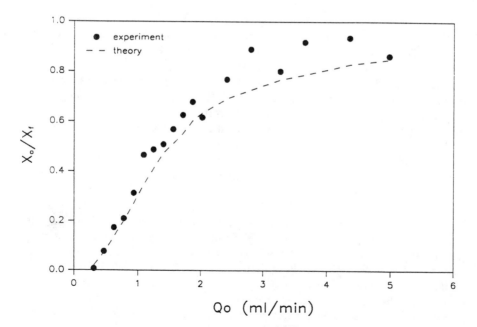

Figure 3.   Relative total cell mass concentration in the settler overflow as a function of the overflow rate for a mixed culture of dividing and nondividing yeast cells.

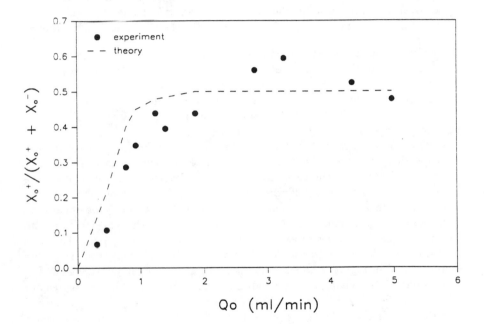

Figure 4.   Fraction of plasmid-bearing, nondividing yeast cells in the settler overflow as a function of the overflow rate for a mixed culture of dividing and nondividing yeast cells.

In this figure, the (+) superscript refers to the nondividing strain, and the (−) superscript refers to the dividing strain. At low overflow rates, the larger cells had sufficient time to settle out of suspension and so the overflow contained primarily dividing cells. As the overflow rate was increased, a greater percentage of the larger, nondividing cells reached the settler overflow so that the selectivity of the separation was decreased. At very high overflow rates, the composition of the overflow stream approached the 50:50 composition of the feed stream. There is good agreement between the data and the theoretical prediction obtained by applying equation 10 independently for the two subpopulations.

Further experiments were performed with separate cultures of just dividing cells or just nondividing cells. The measured concentrations and size distributions in the overflow as functions of the overflow rate and the angle of inclination for these experiments are also in good agreement with the predictions of the theory.

Selective Separation of Hybridoma Cells. In Figure 5, the ratio of cell number concentration in the overflow to that in the bioreactor is shown as a function of the overflow rate from the settler for both the viable and nonviable subpopulation of hybridoma cells. The chemostat culture was operated at a dilution rate of 0.89 day^{-1} and had a total cell number concentration of 2.5×10^6 cells/ml, with a viable fraction of 92% in the bioreactor. At overflow rates less than 20 ml/hr, both the viable and nonviable cells had sufficient time to settle out of suspension before reaching the overflow. As the overflow rate was increased, the smaller, nonviable cells began to appear in the overflow stream, whereas the larger, viable cells still had sufficient time to settle. When the overflow rate was in the approximate range 50 ml/hr $< Q_o <$ 125 ml/hr, a very good separation occurred in which a significant fraction of the nonviable cells was partitioned to the fine fraction (overflow stream), whereas nearly all of the viable cells were partitioned to the coarse fraction (underflow stream). This separation is further demonstrated in Figure 6, which contains forward angle light scattering (FALS) histograms of the viable and nonviable cell subpopulations in the bioreactor and the overflow stream, for a perfusion experiment having $Q_o = 72$ ml/hr and a viable fraction of 50% in the bioreactor. The overflow stream contained a negligible ($< 0.1\%$) quantity of viable cells, and its cell size distribution is close to that of the nonviable subpopulation in the bioreactor. As the overflow rate was increased further, the viable cells also began to appear in significant amounts in the overflow, and the resolution of the separation decreased.

Further experiments were performed with a longer settler ($L = 37$ cm) using a culture that contained a total of 3.9×10^6 cells/ml in the bioreactor, with a viable fraction of 60%. The lower dilution rate for this set of experiments, 0.39 day^{-1}, caused a lower growth rate and, hence, a slightly smaller size of the viable cells than for the experiments with a dilution rate of 0.89 day^{-1}. Nevertheless, the difference in sizes between the viable and nonviable subpopulations was sufficient for a significant separation to be achieved. This may be seen in Figure 7, where the viable fraction and the mean cell diameter in the overflow stream are plotted as functions of the overflow rate. Also shown is the mean cell

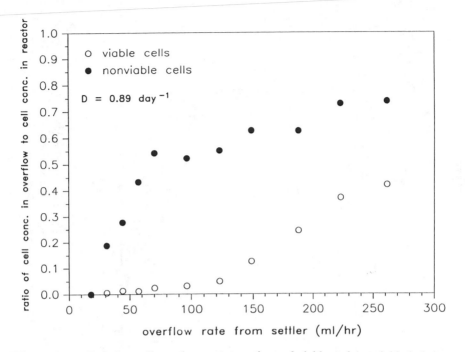

Figure 5.   Relative cell number concentrations of viable and nonviable hybridoma cells in the settler overflow as a function of the overflow rate.

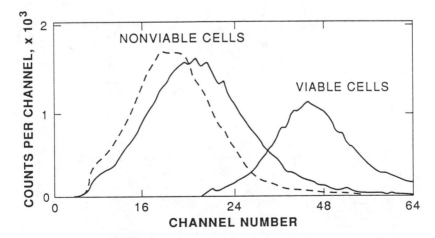

Figure 6.   FALS histograms of viable and nonviable hybridoma cell subpopulations in bioreactor (solid lines) and overflow (dashed line). The number of viable cells reaching the overflow stream was too small to be shown on the scale of this graph.

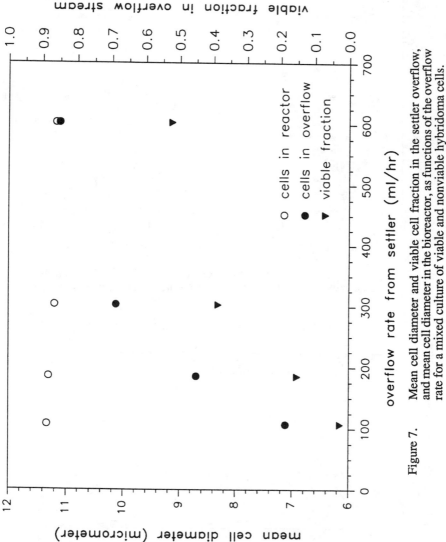

Figure 7.   Mean cell diameter and viable cell fraction in the settler overflow, and mean cell diameter in the bioreactor, as functions of the overflow rate for a mixed culture of viable and nonviable hybridoma cells.

diameter in the bioreactor, which declined only slightly from 11.3 μm to 11.2 μm over the course of the experiments. At low overflow rates ($Q_o < 100$ ml/hr), no viable cells reached the overflow, and the mean cell diameter in the overflow was representative of the smaller, nonviable cells. As the overflow rate from the inclined settler was increased, larger cells were able to reach the overflow without settling out of suspension, and the fraction of viable cells in the overflow stream increased. At very high overflow rates, neither the nonviable or viable cells had sufficient time to settle, and both the viable cell fraction and the mean cell diameter in the overflow approached the corresponding values in the feed stream from the bioreactor.

Concluding Remarks

In this chapter, we have described experiments that demonstrate the ability of a continuous inclined settler system to selectively separate subpopulations of nonflocculent yeast and hybridoma cells on the basis of differences in sizes and sedimentation velocities. The separation is optimum at intermediate overflow rates chosen such that a significant fraction of the smaller subpopulation is partitioned to the overflow stream whereas nearly the entire larger subpopulation is partitioned to the underflow stream. The experimental results for the separation of dividing and nondividing yeast cells are in good, quantitative agreement with the predictions of theory. The results for the separation of viable and nonviable hybridoma cells are in qualitative agreement with the theory, but quantitative comparisons will not be possible until we measure the densities of the viable and nonviable cells and we obtain a sorting capability on the flow cytometer so that accurate size distributions of completely separated subpopulations may be obtained.

Inclined settlers are especially suitable for high capacity separations of cell subpopulations because of their simple construction and their ability to be operated on a continuous basis without any moving parts. Moreover, they may be scaled up directly by increasing the area available for settling. This may be accomplished by using longer or wider channels, or by placing several closely–spaced inclined plates in a larger settling tank in order to obtain multiple channels in parallel. Inclined settlers operate at much higher capacities than vertical settlers because the cells must settle only a distance of order b in an inclined settler, rather than a distance of order L as in a vertical settler. Aspect ratios as large as $L/b = 100$ have been used in practice, resulting in a greater than 50–fold increase in the separation rate for an inclined settler relative to a corresponding vertical settler (*23,24*).

As discussed previously, selective cell separations using inclined settlers may be of practical importance in the biotechnology industry because productive cells may be selectively separated and recycled to a continuous bioreactor while nonproductive cells are removed (*16–18*). We have begun to investigate this strategy using the yeast and hybridoma cells described in this chapter. Our preliminary results with the yeast cultures show that high cell concentration (10^9 cells/ml) and high foreign protein productivities may be maintained for one or more weeks with cell recycle using an inclined settler, whereas the plasmid–bearing cells are washed out of the bioreactor, and the foreign production consequently halted, after only two days when no recycle is used. Similarly, in a perfusion experiment using an inclined settler designed to recycle more than 99.9% of the viable cells to the culture vessel, and to remove a significant fraction of the nonviable cells, we were able to maintain high viable cell concentrations (10^7 cells/ml) and high monoclonal antibody productivities (50 μg/ml–day) over the two–week duration of the experiment.

Acknowledgments

We wish to acknowledge grants EET–8611305 and BCS–8912259 from the National Science Foundation in support of the yeast separations, and grants BCS–8857719 from the National Science Foundation and NB9RAH90H130 from the Department of Commerce in support of the hybridoma separations.

Literature Cited

1. Pretlow, T.G., II; Pretlow, T.P. In *Cell Separation Science and Technology*; Kompala, D.S.; Todd, P., Eds.; ACS Symposium Series, 1991, this volume.
2. Keng, P.C. In *Cell Separation Science and Technology*; Kompala, D.S.; Todd, P., Eds.; ACS Symposium Series, 1991, this volume.
3. Van Alstine, J.M.; Snyder, R.S.; Karr, L.J.; Harris, J. *J. Liquid Chromatog.* **1985**, *8*, 2293.
4. Fisher, D.; Walter, H. In *Cell Separation and Science Technology*; Kompala, D.S.; Todd, P., Eds.; ACS Symposium Series, 1991, this volume.
5. Bigelow, J.C.; Nabeshima, Y.; Kataoka, K.; Giddings, J.C. In *Cell Separation Science and Technology*; Kompala, D.S.; Todd, P., Eds.; ACS Symposium Series, 1991, this volume.
6. Kataoka, K.; Kadowaki, T.; Nabeshima, Y.; Tsuruta, T.; Sakurai, Y. In *Cell Separation Science and Technology*; Kompala, D.S.; Todd, P., Eds.; ACS Symposium Series, 1991, this volume.
7. Miltenyi, S.; Mueller, W.; Weichel, W.; Radbruch, A. In *Cell Separation Science and Technology*; Kompala, D.S.; Todd, P., Eds.; ACS Symposium Series, 1991, this volume.
8. Powers, F.; Heath, C.A.; Ball, E.D.; Vredenburg, J.; Converse, A.O. In *Cell Separation Science and Technology*; Kompala, D.S.; Todd, P., Eds.; ACS Symposium Series, 1991, this volume.
9. Bauer, J. In *Cell Separation Science and Technology*; Kompala, D.S.; Todd, P., Eds.; ACS Symposium Series, 1991, this volume.
10. Todd, P. In *Cell Separation Science and Technology*; Kompala, D.S.; Todd, P., Eds.; ACS Symposium Series, 1991, this volume.
11. Giddings, J.C.; Barman, B.N.; Zhang, J.; Liu, M.-K. In *Cell Separation Science and Technology*; Kompala, D.S.; Todd, P., Eds.; ACS Symposium Series, 1991, this volume.

12. Leary, T.F. In *Cell Separation Science and Technology*; Kompala, D.S.; Todd, P., Eds.; ACS Symposium Series, 1991, this volume.

13. Buican, T. In *Cell Separation Science and Technology*; Kompala, D.S.; Todd, P., Eds.; ACS Symposium Series, 1991, this volume.

14. *Cell Separation—Methods and Selected Applications*; Pretlow, T.G.; Pretlow, T.P., Eds.; Academic Press: 1987, Vol. 1-5.

15. Davis, R.H.; Hunt, T.P. *Biotech. Progress* **1986**, *2*, 91.

16. Davis, R.H.; Parnham, C.S. *Biotech. Bioeng.* **1989**, *33*, 767.

17. K.L.; Davis, R.H.; Taylor, A.L. *Biotech. Progress* **1990**, *6*, 7.

18. Ogden, K.L.; Davis, R.H. *Biotech. Bioeng.* **1991**, *37*, 325.

19. Davis, R.H.; Zhang, X.; Agarwala, J.P. *Ind. Eng. Chem. Res.* **1989**, *28*, 785.

20. Davis, R.H.; Acrivos, A. *Ann. Rev. Fluid Mech.* **1985**, *17*, 91.

21. Szlag, D.C. *Factors Affecting Yeast Flocculation*, **1988**, M.S. Thesis, Univ. Colo.: Boulder, CO

22. Davis, R.H.; Hassen, M.A.; *J. Fluid Mech.* **1988**, *196*, 107.

23. Davis, R.H.; Herbolzheimer, E.; Acrivos, A. *Phys. Fluids* **1983**, *26*, 2055.

24. Davis, R.H.; Birdsell, S. *Dev. Ind. Microbio.* **1985**, *26*, 627.

RECEIVED March 15, 1991

Chapter 9

Separation of Cells by Field-Flow Fractionation

J. Calvin Giddings[1], Bhajendra N. Barman[2], and Min-Kuang Liu[1]

[1]Field-Flow Fractionation Research Center, Department of Chemistry,
University of Utah, Salt Lake City, UT 84112
[2]FFFractionation, Inc., P.O. Box 58718, Salt Lake City, UT 84158

Since the development of steric field-flow fractionation
(steric FFF) in our laboratories over ten years ago,
application to biological cells has been an attractive
prospect. While our earlier developments in FFF were
primarily applicable to macromolecules and submicron
particles, the development of steric FFF, and later
hyperlayer FFF, extended the range of FFF operation to cell-
sized (1-100 μm) particles.
    The principles of steric and hyperlayer FFF are
briefly reviewed here and their separative capabilities
demonstrated using latex microspheres. Several applications to biological cells, particularly blood cells, are
reported. Some of the unique operating characteristics of
these FFF techniques for cell separations are summarized.
These characteristics include high resolution, rapid
separation, simplicity of operation, and an ability to
characterize as well as separate cell populations. We also
note a relatively new modification of FFF capable of
providing continuous separation.

Field-flow fractionation (FFF) is a relatively new and powerful family of
methods having broad capabilities in the separation of macromolecules
and particles (1-5). Application to biological cells is a natural outgrowth
of the ongoing development of FFF technology.
    FFF is a chromatographic-like technique with separation occurring
in a thin unobstructed flow channel resembling that depicted in Figure 1.
A volume of sample is injected into the inflowing carrier stream and separated in the FFF channel. Separation is followed by elution, detection, and
fraction collection. A typical elution profile (fractogram) of cell-sized
particles of different diameters is shown in Figure 2. (This separation was
obtained using FFF channel I, described later.) Large particles generally
emerge before small particles in this size range as illustrated in the
figure. The size range can be expanded in either direction (presently

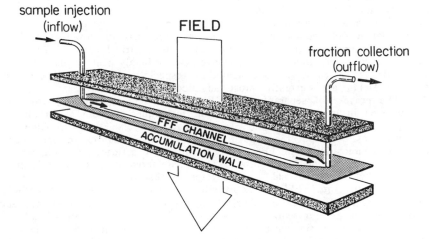

Figure 1. Exploded view of FFF channel cut from a plastic spacer and sandwiched between two wall elements.

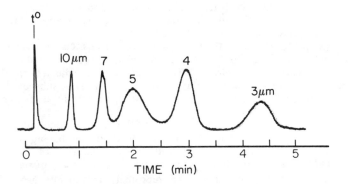

Figure 2. Separation of polystyrene latex particles of indicated diameters using crossflow as a driving force. The aqueous carrier contained 0.1% (v/v) of FL-70 detergent and 0.01% (w/v) sodium azide. Channel and cross flowrates were 5.17 and 1.12 mL/min, respectively.

realized size range of FFF: 0.002-500 µm) by changing experimental conditions.

FFF is a batch process in which discrete samples are separated (repetitively if desired). A similar channel configuration differing only at the ends where flow splitters are inserted can be used for the continuous separation of cells or other particles. These channels are known as split-flow thin (SPLITT) cells (6-8).

Differential retention and thus separation in FFF systems is induced by an external field acting in a direction perpendicular to flow. The field may be as common as gravity or as powerful as electrical forces. The only limitation is that the field must be convenient to apply and it must generate the desired selectivity. Sedimentation (especially in a centrifuge) and crossflow fields are currently most advanced experimentally.

The process of separating cells by FFF differs substantially from that   utilized in most other cell separation techniques. There are a number of advantages inherent to the FFF approach but also a few problems and limitations. To put these positive and negative features in perspective, the major characteristics of FFF that determine its efficacy for cell separations are briefly summarized below. The characteristics noted in this list will be further discussed and detailed later in this paper.

1. <u>Resolution</u>. FFF yields excellent resolution for cell-sized particles based on small differences in physical properties such as size, charge, and density. Based on our work with polystyrene (PS) latex spheres--ideal probes because they are narrower in distribution than most cell populations--particles differing in effective diameter by only 20-50% can be fully resolved from one another by FFF.

2. <u>Sample handling</u>. The samples injected into the carrier stream entering an FFF channel are divided into components that emerge at different times and therefore can be simply collected as fractions from the outlet flowstream. There is little manual effort needed in sample handling and collection. The entire procedure is subject to automation through the use of an autoinjector and a fraction collector.

3. <u>Separation speed</u>. Most cellular suspensions can be separated into component cell populations and the components can be collected in 2-8 minutes using current technology. This time can be adjusted to fit special needs. In fractograms displayed in Figure 2 and subsequent figures, cells and particles down to 3 µm elute generally in less than 5 minutes.

4. <u>Capacity and throughput</u>. Each FFF run, depending upon channel dimensions (typically 0.025 cm × 2 cm × 30 cm, having a volume of 1.5 mL or so), can accommodate 1-30 million cells the size of erythrocytes without an undue loss of resolution (see later). Fewer cells of larger size will be separable. If the method is used repeatedly to separate 5×10^6 cells with a cycle time of three minutes, then 10^8 cells could be separated in one hour. To enhance the throughput substantially above this level it would be necessary to use continuous SPLITT cell methodology as noted earlier (6-8).

5. <u>Gentleness</u>. The applied field is expected to have very little disruptive effect on cells. The short duration of the separation process helps maintain cell integrity. However, while flowing in the channel, cells are subject to a moderate (but adjustable) shear rate which typically ranges from 100 to 2000 s^{-1}. This shear will increase briefly but substantially at the ends of the FFF channel, a problem needing additional atten-

tion. On the other hand, the potentially disruptive effects of repeatedly sedimenting and resuspending cell populations is avoided in FFF.

6. Adhesion to apparatus walls. The shear processes noted above can be adjusted to generate hydrodynamic lift forces strong enough to prevent particle contact with the channel wall. By using new injection procedures (utilizing hydrodynamic relaxation) developed in our laboratory (9, 10), it should be possible to run cells through the entire separation cycle without cell-to-wall contact.

7. Suspending media. In most cases any desired suspending medium can be chosen for separation, giving an environment that is most favorable to cell integrity and longevity. Only in special cases is a density enhanced medium needed for cell separations. When density enhancement is necessary, the solution densities can often be kept below that of the cells.

8. Impurities. Low molecular weight impurities, unwanted cells, and miscellaneous debris can generally be eluted from the channel at times different from the elution times of desired cell populations. Thus cells are not only separated from each other but (in the same step) from most impurities. Sample preparation may consequently be simplified.

9. Flexibility. FFF is a flexible methodology that can be customized to fit the needs of different cell separation problems. Different fields can be used to change the basis of selectivity, some options being separation based on effective size (flow FFF), size and density (sedimentation FFF), and charge (electrical FFF). The field strength can be controlled to vary the retention level and to change the range of applicability. Variations in flow velocity can be used to govern separation speed and control shear effects (1-5).

10. Cell characterization. The fractogram (detector signal versus time) of cell samples provides a rapid means of characterizing the size distribution of cells, relative population levels, non-cellular impurities, and other features of the sample population. Such fractograms can be readily computerized for rapid data handling and analysis.

11. Simplicity of Apparatus. The FFF apparatus is intrinsically simple in physical structure and in operation. Consequently, the cost of such apparatus and the cost of operation is relatively low.

Despite the promise shown by FFF for cell separations, little work has been reported in this area. In 1984 Caldwell et al. first showed the applicability of FFF to human and animal cells using centrifugal forces (11). Figure 3 (taken from that study) shows the separation of freshly trypsinized HeLa cells from fixed erythrocytes. The HeLa cells were shown to retain viability after passage through the sedimentation FFF system.

Martin and his colleagues have reported on the separation of various blood cell populations using FFF with gravity as the driving force (12). Unpublished studies carried out in our laboratories have shown that the size distribution of erythrocytes can be readily obtained in a run time of a few minutes using a crossflow driving force. Other unpublished work has shown that lymphocytes tend to stick to the membrane wall when crossflow is used. We believe that this problem can now be circumvented using hydrodynamic relaxation.

We have recently reported another approach--a hybrid of FFF and adhesion chromatography--for the separation of B and T cells, as described in the next chapter in this volume (13).

Mechanism of Separation

The mechanism by which cell-sized particles are separated in FFF is illustrated in Figure 4. (A different mechanism is applicable to macromolecules and particles of submicron size (1-5).) According to this mechanism, particles are driven by the applied field toward one wall, termed the accumulation wall. They approach an equilibrium position in which the driving force is exactly balanced by repulsive forces originating at the wall.

The most important forces of repulsion in FFF are hydrodynamic lift forces (14). These forces are usually attributed to inertial effects, which have been investigated extensively (15-19). However, recent studies suggest that forces related to lubrication phenomena play a major role in driving cell-sized particles away from the accumulation wall (20).

Once in an equilibrium position, the particle is swept down the channel by the flowing carrier liquid. The flow velocity in the channel is differential so the velocity of the particle depends upon its equilibrium position. Specifically, the flow in all practical cases is laminar and the resulting flow profile is parabolic (Figure 4). A particle whose equilibrium position lies close to the accumulation wall of the channel is displaced only slowly by flow because the flow velocity approaches zero at the wall. A particle with an equilibrium position further removed from the wall is caught up in higher velocity streamlines and is displaced more rapidly. Thus particles with different equilibrium positions are separated because they are displaced by flow at different velocities.

At low flowrates each particle approaches contact with the wall before reaching equilibrium. In such a position, the particle will travel at a velocity approximately equal to the flow velocity at its center of gravity, which will be located a distance only slightly greater than particle radius a from the wall. Under such conditions, larger particles, whose centers are farther removed from the wall than those of smaller particles by bulk steric effects, are displaced more rapidly than the smaller particles. The achievement of separation based on such size-related steric effects is termed *steric FFF*.

In many cases the mean flow velocity <v> is increased to values greater than 1 cm/s in order to hasten the completion of separation. Flow velocities of this magnitude in typical channels of thickness 0.2-0.3 mm lead to substantial hydrodynamic lift forces that tend to drive the particles away from the wall (14). Thus at a given field strength the distance δ between the nearest particle surface and the accumulation wall increases with flow velocity. Clearly the values of δ can be manipulated by increasing the flowrate (which makes δ larger) or by increasing the field strength (which makes δ smaller).

As δ values increase and finally exceed the particle radius a, the FFF methodology is labeled *hyperlayer FFF*. For reasons too lengthy to describe here, a sedimentation field is generally most effective when utilized in the steric mode of operation, giving the subtechnique termed *sedimentation/steric FFF*. When crossflow is used as the driving force, the hyperlayer mode generally produces the best results (14). The subtechnique is then labeled *flow/hyperlayer FFF*. Most of the ensuing discussion will focus on these two subtechniques of FFF.

An important aspect of FFF operation is the relaxation process in which particles entering the channel, generally spread out over its full

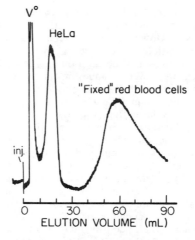

Figure 3. Separation of HeLa from fixed erythrocytes using a centrifugal driving force. Symbol V^0 indicates the void volume of the channel. (Reproduced with permission from reference 11. Copyright 1984 Humana Press.)

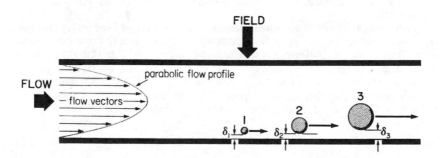

Figure 4. Edge view of FFF channel showing the equilibrium positions and relative velocities of three cell-sized particles in the parabolic channel flow.

cross section, are driven to their respective equilibrium positions. The time required for the relaxation of cell-sized particles varies from a few seconds to a few mintues.

If flow continues unabated during this relaxation process (*stopless flow* operation), identical particles will be scattered up and down the channel depending upon their original cross sectional position. Thus resolution will be lost. The traditional means for avoiding this problem in FFF is to stop the flow (*stopflow* operation) shortly after sample injection so that the particles can reach or approach their equilibrium positions under static conditions. However with interrupted flow, lift forces temporarily vanish and the particles have access to the surface, risking adhesion. This may or may not be a problem depending upon the nature of the cellular components. When adhesion is a problem, the risk can likely be reduced or eliminated by reducing (but not eliminating) flow during sample introduction (*slow flow* operation). However the most promising technique for eliminating the risk of adhesion during the vulnerable period of relaxation is that in which particles are subjected to hydrodynamic relaxation upon entering the FFF channel.

Several means for achieving hydrodynamic relaxation have been developed recently in our laboratory (9, 10). In general, hydrodynamic relaxation is realized by introducing two separate substreams of liquid into the channel. The larger stream, free of particles, enters as a lamina that compresses the particle-containing stream, also entering as a lamina, into a thin layer adjacent to the accumulation wall. Thus by flow alone, particles are driven near their equilibrium positions; the small displacements needed to achieve final equilibrium are nondisruptive to the separation process. Means for introducing the two laminar streams into the channel include the utilization of a thin flow splitter bisecting the channel across its thin dimension (9). In this case the inlet configuration resembles that of continuous SPLITT cells. A second means utilizes a small frit element embedded in one channel wall through which the particle-free stream is introduced (10). This method is used later in this paper (channel II).

We noted above that when particles are separated according to the steric mechanism, larger particles, protruding more deeply into the flow stream than smaller particles, are displaced more rapidly through the channel than the smaller particles. The velocity V_p of any given particle population relative to the mean fluid velocity <v> is defined as the retention ratio R for that particle type. The magnitude of R is related to the particle radius a or diameter d by the equation

$$R = \frac{6\gamma a}{w} = \frac{3\gamma d}{w} \tag{1}$$

where w is the channel thickness and γ is a dimensionless steric correction factor ($\gamma \cong 1 + \delta/a$) that accounts for the interplay of lift forces and driving forces. The exact calculation of γ is presently difficult because of the complexity of lift forces (20). We note that radius a and diameter d must be interpreted as effective radii and diameters for nonspherical particles.

Retention is conveniently described in terms of particle retention time t_r. Since $R = t^0/t_r$, where t^0 is the mean residence time of fluid in the channel, t_r is given by

$$t_r = \frac{\ell\, w}{6\gamma a} = \frac{\ell\, w}{3\gamma d} \tag{2}$$

We observe that Equations 1 and 2 predict a high level of selectivity in the separation of particles of different size. For example, if radius a (or diameter d) differs by 20% from one particle to the next, the values of R and t_r should likewise differ by 20%, which is often sufficient for complete separation. However this conclusion is based on the assumption of a constant γ, which does not hold in practice. The value of γ, strongly influenced by lift forces, depends upon particle diameter, flow velocity, and driving force (20). Because γ depends on d, the selectivity may vary above or below the value noted above. Specifically, if we define the selectivity as

$$S_d = \frac{percent\ shift\ in\ t_r}{percent\ difference\ in\ d} \tag{3}$$

then when γ is constant, as described in the example above, $S_d = 1$. However the value of S_d in sedimentation/steric FFF, once γ is accounted for, lies in the approximate range 0.5-0.8. This value is somewhat higher in flow/hyperlayer FFF where selectivities up to 1.5 are often realized (*14*).

Equations 1 and 2 suggest that particles are separated only because of differences in size. However, since γ is influenced by the driving force, it is also influenced by particle density in the case of sedimentation/steric FFF. (For all forms of flow FFF, retention is independent of density.) It has been shown that particles differing significantly in density but not in size can be separated using sedimentation/steric FFF (*21*). This effect would be enhanced by increasing the carrier density to a value approaching that of the least dense particle. We note that if the carrier density lies between that of the two particles, they will undergo FFF migration at opposite walls. The two populations can then be separated by an outlet splitter if desired (*22*).

Experimental

We have used two flow FFF systems to further characterize parameters relevant to cell separations. Both walls of these channels are made of ceramic frit to make them permeable. The lower (accumulation) wall is covered with a membrane to allow the passage of carrier fluid but retard entrained particles.

Channel I is a normal flow FFF system resembling that shown in Figure 1. The dimensions of Channel I are tip-to-tip length $L = 27.8$ cm, breadth $b = 2.0$ cm, and thickness $w = 254$ μm. A PM10 membrane from Amicon (Danvers, MA) was used at the accumulation wall.

Channel I was coupled with a Constametric I pump from LDC Analytical (Riviera Beach, FL) to deliver the channel inflow. An Isochrom LC pump (Spectra-Physics, San Jose, CA) was used to deliver the inlet crossflow. The outlet crossflow was controlled by a Minipuls II peristaltic pump (Gilson, Middleton, WI) used as an unpump. A Model 7125 valve from Rheodyne (Cotati, CA) equipped with a 20 μL sample loop was used for the sample injection. Both polystyrene (PS) beads and blood cells were detected at 350 nm with a Model 757 spectrophotometric detector from Applied Biosystems (Ramsey, NJ). A strip chart recorder from Houston Instruments (Austin, TX) was used to monitor detector response.

Channel II is similar in many respects to Channel I but provisions have been made to carry out hydrodynamic relaxation. This is accomplished by using a frit inlet system in which a small element of frit near the inlet is subjected to a separately controlled crossflow that drives the sample inlet substream into the thin lamina desired (10). The dimensions of Channel II are $L = 38.0$ cm, $b = 2.0$ cm, and $w = 220$ μm. The membrane used here is a YM10 membrane, also from Amicon.

A Cheminert metering pump (Chromatronix, Berkeley, CA) was used to deliver the sample stream (flowrate $\dot{V}_s$) into channel II. A Model 414 LC pump from Kontron Electrolab (London, England) was used for the frit inlet stream (flowrate $\dot{V}_f$). These two flows combined yield the channel flowrate $\dot{V} = \dot{V}_s + \dot{V}_f$. An in-house built syringe pump was used to deliver the crossflow stream (flowrate $\dot{V}_c$) into Channel II. The outlet channel flow and crossflow were controlled by flow resitrictors made of 0.025 cm internal diameter stainless steel tubing. A Model 7010 Rheodyne injection valve with a 34 μL sample loop was used for sample injection. The eluting sample was detected by a Model 757 spectrophotometric detector from Applied Biosystems and the detector signal was fed to a Model SS-250F recorder (Esterline Angus Instrument, Indianapolis, IN). The detector wavelengths were set at 254 nm for PS beads and 350 nm for blood cell studies.

Isoton II (Coulter Diagnostics, Hialeah, FL) and laboratory prepared phosphate buffered saline, pH 7.3, osmolality 290 mmol/kg, were used as carrier liquids for blood cell separations. Doubly distilled water with 0.1% (v/v) FL-70 (Fisher Scientific, Fair Lawn, NJ) and 0.02% (w/v) sodium azide (Sigma, St. Louis, MO) was used for the separation of polystyrene latex beads. All experiments were carried out at room temperature. The channels in both systems were placed with their principal axis in a vertical position to avoid gravitational effects on the transverse displacement of particles or cells.

Results and Discussion

Channel I Studies. Channel I was used to examine the characteristics and the FFF behavior of red blood cells (RBCs) from various mammals. Fresh untreated blood was used as the sample material. The whole blood was diluted 100-fold with the appropriate carrier liquid before injection from a completely filled 20 μL sample loop. Therefore, 1 to 2 million RBCs were injected for each run as established by RBC counts obtained from a Coulter Counter. The detector signal was due almost entirely to the erythrocyte content of the blood.

The results clearly show a differential retention for erythrocytes of different origin. Thus in Figure 5 we show the significantly different fractograms obtained for dog, human, horse, and cat erythrocytes in the phosphate buffer. A fractogram of four latex standards is shown for reference.   For these runs the channel and cross flowrates ($\dot{V}$ and $\dot{V}_c$) were 5.0 and 1.3 mL/min, respectively.   A stopflow time of 1.5 min was employed.

The observed mean retention times for the erythrocytes in Figure 5 follow the sequence: dog < human < cat $\leq$ horse. (This elution is not the same as observed for the Coulter Counter, perhaps because of conformational changes caused by the phosphate buffer. The two sequences are in accord for Isoton II buffer.) The size distribution of the cells is clearly broader than that of the latex standards. These results can be very accurately reproduced (to within about one percent) from one run to the next. Different features (e.g., average, mode, and variance) of the size distribution for each cell population can be readily deduced from the fractograms.

The RBC counts and the effective spherical diameter d_{cc} based on the mean cell volume (MCV) determined by the Coulter Counter together with the mean effective diameter d_{FFF} determined by the flow/hyperlayer FFF measurements shown in Figure 5 are summarized in Table I. The standard deviation (s.d.) in effective cell diameters for the different RBC samples of Figure 5 are also provided in the table. The good agreement between Coulter and FFF diameters (d_{cc} and d_{FFF}) is fortuitous because the former is an effective spherical diameter and the latter an effective hydrodynamic diameter.

Table I.   Coulter Counter and Fl/HyFFF Results for the Mammalian
Red Blood Cell Samples Utilized in Figure 5

	Coulter Counter Results			Fl/HyFFF Results
	RBC counts (10^6/µL)	MCV (fL)	d_{cc} (µm)	d_{FFF} ± s.d. (µm)
Dog	7.72	71.6	5.15	6.11 ± 1.07
Human	4.44	91.9	5.60	5.64 ± 0.97
Horse	8.29	51.0	4.60	4.69 ± 0.65
Cat	6.74	53.0	4.66	4.78 ± 0.53

Figure 6 shows the elution pattern for erythrocytes obtained from human and bovine blood using Isoton II as the carrier liquid. The flowrates were $\dot{V}$ = 5.2 and $\dot{V}_c$ = 1.9 mL/min. These two erythrocyte populations, when combined, are clearly separated into distinct peaks by FFF, as shown in the lower fractogram.

Channel II Studies.   Channel II was used to better characterize the capacity of FFF systems for cells. Initial results were obtained using two well defined polystyrene latex standards (15 µm and 10 µm in diameter) that are normally well resolved from one another (see Figure 7a). The mixture of the two standards was injected into the channel by means of the injection valve and 34 µL sample loop described earlier. Different

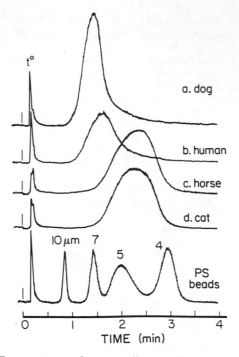

Figure 5.   Fractograms of mammalian erythrocytes with four latex standards as reference.

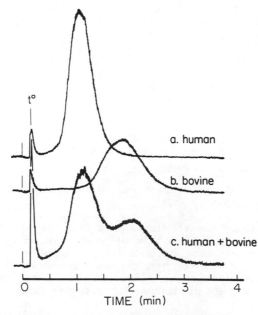

Figure 6.   Fractograms of human and bovine blood and a mixture of the two.

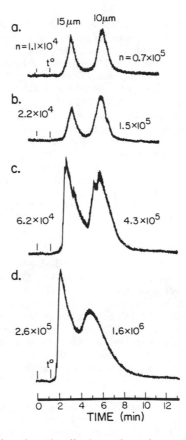

Figure 7.   Capacity (overloading) study using two cell-sized latex populations having sphere diameters of 15 and 10 μm.   The number (*n*) of injected particles is shown next to each peak in the series of fractograms.

concentrations were used in succeeding runs. The results of four of the runs are shown in Figure 7. As the particulate content increases (from top to bottom), the resolution of the two populations gradually erodes. The numbers of injected particles (n), as indicated at each peak in the figure, were obtained by a standard cell counting procedure using a Neubauer hemacytometer and were in good agreement with the amounts calculated based on the latex concentration provided by the supplier.

The results of Figure 7 suggest that it is difficult to resolve, in a single run, much more than 10^6 particles (cells) 10 μm or more in diameter. This number should increase for smaller cells and for cells of greater relative size difference and decrease for larger cells and for populations closer to one another in relative size. Other factors, presently unkown, may influence capacity; virtually no previous work has been done in this area on particles the size of cells.

It is interesting to compare these results with those observed for red blood cells. In the studies with channel I, no distortion (overloading) of the erythrocyte fractograms was observed for samples of up to 5×10^6 cells. If overloading is strictly a volumetric effect, then the capacity for erythrocytes, whose volumes are typically 90 femtoliters for human cells, should be approximately six times higher (in cell numbers) than the capacity for 10 μm spheres and 20 times higher than the capacity for 15 μm spheres. The hypothesis that capacity is determined by total particulate volume is roughly in accord with our experience with erythrocytes and with Figure 7 except there may be some tendency for the latex spheres to overload more readily than the erythrocytes.

To further characterize capacity, we carried out additional studies in channel II using a far wider range of erythrocyte concentrations. The conditions for this series of runs were $\dot{V} = 3.45$ and $\dot{V}_c = 1.93$ mL/min. The ratio of frit to sample inlet flowrates was 94/6. The blood sample was suspended in Isoton II without further treatment. Figure 8 shows the superposition of fractograms of two human blood samples, one having 1.6 million cells and the other 33 million cells, as determined from a cell count derived from a Coulter Counter. We note that the peak areas are not proportional to the cell count because of different response settings. We observe that the 20-fold increase from 1.6 to 33 million cells has the effect of tilting the peak somewhat forward and broadening it out. However the broadening does not appear to be great enough to prevent the separation of erythrocytes from other cells.

In order to test the possibility of separating cells in the presence of so many erythrocytes, a run of the 33 million cell sample was made under the same conditions as described above except in this case fractions were collected at different times. Figure 9 shows the detector signal for this run and the times at which fractions were collected. The fractions were examined by microscopy to establish cell identity. The micrographs presented in Figure 9 show that a population of white cells precedes the elution of the main body of red cells. (The identity of the larger bodies as white cells was confirmed by a staining procedure.) While these results are very preliminary, we believe that the procedure can be improved even further to yield clean populations of the two types of cells.

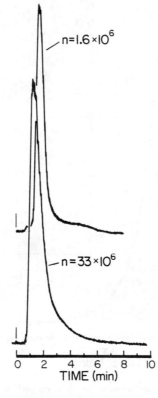

Figure 8. Effect of number (n) of erythrocytes on elution profile from Channel II.

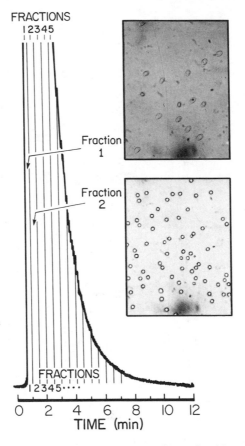

Figure 9.   Micrographs confirming separation of white and red cells, which are found in fractions 1 and 2, respectively.   The time intervals of collection of these and other fractions are shown on the accompanying   fractogram.

Conclusions

Field-flow fractionation is a chromatographic-like family of techniques having many intrinsic advantages in the separation of cells and other biological particles.   However, the FFF methods are not widely known in the biological community and thus very little work has been previously reported on the separation of cells by FFF.   In this chapter, in addition to outlining the principles of FFF as they apply to cell-sized bodies, we have presented preliminary results verifying the high speed and effective separation and characterization of several cell populations.

Legend of Symbols

a	particle radius
b	channel breadth
d	particle diameter
L	channel length
n	number of injected particles
R	retention ratio
S_d	selectivity
t^0	mean residence time of fluid in channel
t_r	particle retention time
$<v>$	mean flow velocity
$\dot{V}$	channel flowrate
$\dot{V}_c$	cross flowrate
$\dot{V}_f$	frit inlet flowrate
V_p	velocity of particle population
$\dot{V}_s$	sample stream flowrate
w	channel thickness

Greek

δ	distance between particle surface and accumulation wall
γ	dimensionless steric correction factor

Acknowledgments

The authors thank Dr. Ed Ashwood for supplying most of the human and animal cells used in this study.   This work was supported by U.S. Public Health Service Grants GM10851-33 and 1 R43 HL40724-01A1 from the National Institutes of Health.

Literature   Cited

1.   Giddings, J. C.; Myers, M. N.; Caldwell, K. D.; Fisher, S. R. In *Methods of Biochemical Analysis*; Glick, D., Ed.; John Wiley: New York, 1980; Vol. 26, pp 79-136.
2.   Giddings, J. C. *Anal. Chem.* **1981**, *53*, 1170A.
3.   Giddings, J. C. *Sep. Sci. Technol.* **1984**, *19*, 831.
4.   Caldwell, K. D. *Anal. Chem.* **1988**, *60*, 959A.
5.   Giddings, J. C. *C&E News* **1988**, *66*, 34.

6. Giddings, J. C. *Sep. Sci. Technol.* **1985**, *20*, 749.
7. Springston, S. R.; Myers, M. N.; Giddings, J. C. *Anal. Chem.* **1987**, *59*, 344.
8. Levin, S.; Giddings, J. C. *J. Chem. Tech. Biotechnol.* **1990**, *50*, 43.
9. Lee, S.; Myers, M. N.; Giddings, J. C. *Anal. Chem.* **1989**, *61*, 2439.
10. Giddings, J. C. *Anal. Chem.* **1990**, *62*, 2306.
11. Caldwell, K. D.; Cheng, Z.-Q.; Hradecky, P.; Giddings, J. C. *Cell Biophys.* **1984**, *6*, 233.
12. Cardot, P.; Martin, M. First Int. Symp. on Field-Flow Fractionation, Park City, Utah, June 16, 1989.
13. Bigelow, J. C.; Giddings, J. C.; Nabeshima, Y.; Tsuruta, T.; Kataoka, K.; Okano, K.;Yui, N.; Sakurai, Y. *J. Immunological Methods* **1989**, *117*, 289.
14. Giddings, J. C.; Chen, X.; Wahlund, K.-G.; Myers, M. N. *Anal. Chem.* **1987**, *59*, 1957.
15. Segré, G.; Silberberg, A. *Nature* **1961**, *189*, 209.
16. Segré, G.; Silberberg, A. *J. Fluid Mech.* **1962**, *14*, 136.
17. Saffman, P. G. *J. Fluid Mech.* **1965**, *22*, 385.
18. Cox, R. G.; Brenner, H. *Chem. Eng. Sci.* **1968**, *23*, 147.
19. Ho, B. P.; Leal, L. G. *J. Fluid Mech.* **1974**, *65*, 365.
20. Williams, P. S.; Koch, T.; Giddings, J. C. *J. Fluid Mech.*, submitted.
21. Caldwell, K. D.; Nguyen, T. T.; Myers, M. N.; Giddings, J. C. *Sep. Sci. Technol.* **1979**, *14*, 935.
22. Jones, H. K.; Phelan, K.; Myers, M. N.; Giddings, J. C. *J. Colloid Interface Sci.* **1987**, *120*, 140.

RECEIVED March 15, 1991

AFFINITY ADSORPTION AND
EXTRACTION METHODS

Chapter 10

Separation of Cells and Measurement of Surface Adhesion Forces

Using a Hybrid of Field-Flow Fractionation and Adhesion Chromatography

J. C. Bigelow[1], Y. Nabeshima[2], Kazunori Kataoka[2], and J. Calvin Giddings[3]

[1]Department of Pharmacology, University of Vermont,
Burlington, VT 05405
[2]Department of Materials Science and Research Institute for Biosciences,
Science University of Tokyo, Noda, Chiba 278, Japan
[3]Field-Flow Fractionation Research Center, Department of Chemistry,
University of Utah, Salt Lake City, UT 84112

A hybrid of the techniques of cellular adhesion chromatography (AC) and field-flow fractionation (FFF) has been used for the separation of mammalian cells. This hybrid technique uses FFF-type channels to generate controlled hydrodynamic forces which are sufficient to detach cells from the accumulation wall of the channels. Coatings with well defined adsorption properties (such as those utilized in AC) are used at the accumulation wall to selectively bind cells. These cells are then separated on the basis of different surface adhesion properties.

This technique has been applied to several problems of biological interest. In one application, T and B lymphocytes in a mixture were completely separated from one another using a coating of poly(2-hydroxyethyl methacrylate) /polyamine graft copolymer on the accumulation wall. From the flow velocities necessary to dislodge the cells it could be estimated that the B cells bind with a force five times greater than the T cells. Additionally, the technique can be used to examine adhesion of cultured cancer cells to a variety of surfaces.

Separation of mammalian cells is of considerable importance in scientific research, medicine, and industrial production. This importance is reflected in the wide variety of cell separation methods which have been developed, many of which are which discussed in this volume. As in many areas of separation science, hybrids of techniques and principles can yield methods and information which are new and distinct from the parent technologies. This chapter discusses one such hybrid: namely that of the separation methods of cellular adhesion chromatography (AC) and field-flow fractionation (FFF).

0097–6156/91/0464–0146$06.00/0

Field-flow fractionation is a family of powerful analytical separation methods. These methods have been developed and extensively researched by one of our groups (University of Utah) and have been well demonstrated to be very effective at separating cell-sized particles, colloids, and macromolecules on the basis of mass, size, and other physical properties (See Giddings *et al.* in this volume for a more complete discussion of the applications of FFF to cell separations). Separations by FFF are typically carried out in the interior volume of unpacked thin ribbon-like channels and, unlike chromatographic techniques, there is no stationary phase and no adsorption based separation. In FFF, a carrier liquid is pumped through the channel and assumes a laminar parabolic flow profile in the channel (see Figure 1). A field or gradient (i.e gravitational, centrifugal, thermal) is applied perpendicular to the thin dimension of the channel. Affected species are driven towards a region of low flow near one wall (the accumulation wall) by the field while dispersive forces in turn drive particles back into faster streamlines. Average particle velocity, and thus separation, is the result of the interplay between the dispersive forces, the driving forces, and the parabolic flow profile. Because FFF channels are unpacked, uniform, and geometrically simple, it is possible to easily predict and control the hydrodynamic shear rate resulting from the strong difference in flow velocity between the center of the channel and the channel walls. Hydrodynamic shear is well recognized to be able to detach small particles from surfaces and is a integral part of some modes of FFF (1). This is in sharp contrast to packed column chromatography where local shear rates are essentially unpredictable and may vary greatly.

Cellular adhesion chromatography (AC) is an unrelated cell separation technique. It utilizes physio-chemical based selective adhesion of cell groups to column packing material for separation. As will be reported herein, mammalian cells will adhere very strongly to some common surfaces such as polystyrene or glass but will adhere very weakly, if at all, to simply treated surfaces. Thus the goal of AC is to design surfaces with selectivity and sufficient adhesion, and from which cells can be recovered. Previous work by one of our groups (Science University of Tokyo) with AC has established that a family of poly(2-hydroxyethyl methacrylate)/polyamine graft copolymers (HA copolymers) can be used effectively as a selective column adsorbent surfaces for B and T lymphocyte separations. These copolymers and related polystyrene/polyamine copolymers form a microdomain structure for which B cells have a greater affinity than T cells (2, 3). A column packed with HA coated glass beads selectively adsorbed B cells from a lymphocyte suspension passed through the column and allowed the recovery of more than 60% of the T cells with a purity of at least 90%. The B cell population was more tightly adsorbed and could be recovered only by mechanical agitation (4).

Hybrid FFF/AC combines the controllable hydrodynamic shear forces of FFF and the selective adhesion of AC to yield information and separations which cannot be obtained with either technique alone. In FFF/AC, the accumulation wall of a small FFF-like channel has a well defined surface such as found in AC. Hydrodynamic shear is used to selectively and evenly detach particles from the surface in a controlled fashion. This both enhances the separating ability of AC and also allows cell/surface adhesion to be studied in a systematic fashion. Additionally, control of the hydrodynamic shear also can allow quantitative estimation of the differences in cell/surface adhesion forces.

In this chapter several application of FFF/AC are presented. We demonstrate use of FFF/AC in evaluating cell/surface adhesion characteristics for two cultured cancer cell lines to a variety of surfaces under different conditions. We also discuss the use this hybrid technique in the important separation of T and B lymphocytes subpopulations, a separation which was complete with excellent recovery of both species. Use of FFF/AC also allowed a quantitative estimation T and B cell surface adhesion to be made (5).

Experimental Procedures

Polymer Solution Preparation. A poly(2-hydroxyethyl methacrylate)/polyamine graft copolymer with 13 wt-% polyamine (designated as HA13) was prepared by radical copolymerization of 2-hydroxyethyl methacrylate with polyamine macromonomer. Details of the synthesis are reported elsewhere (4). Poly(2-hydroxyethyl methacrylate) (PHEMA), polyimide, and Nafion (5 wt.-% in a mixture of aliphatic alcohols) were obtained from Aldrich Chemical Company (Milwaukee, WI). HA13 was dissolved in absolute ethanol at 0.2 wt.-%. PHEMA was dissolved in 50%/50% methanol/absolute ethanol at 0.5 wt.-%. Polyimide was dissolved in N,N-dimethlyformamide at 0.5 wt.-%. Nafion, a perfluorinated anion-exchange polymer, was used as received.

Channel Construction. For this study, a small FFF-type channel was constructed by sandwiching a Teflon spacer, with the channel shape cut out, between a top flat glass plate and a bottom slide-glass sized surface with a simple clamping apparatus. The top channel plate consisted of a piece of heavy glass (2.5 cm x 7.6 cm x 0.635 cm). Two ports, one for carrier inlet and one for carrier outlet, were drilled through the top plate 5.1 cm apart and midway across the breadth of the plate. Teflon tubing (0.3 mm I.D.) was force fitted through the holes and the inner plate surface was cut flush. Approximately 0.7 cm of tubing was left protruding from each hole on the top outside surface of the plate for connection of the channel to the flow system. The dead volume of these port was approximately 1 μL. The channel was approximately 5.5 cm long, 0.0254 cm in thickness, 1.0 cm in breadth and had triangular ends cut to accommodate the inlet and outlet ports. The channel volume was approximately 110 μL with a geometric surface area was 4.5 cm^2.

Accumulation Wall Preparation. The bottom channel wall consisted of either bare or polymer coated surfaces. Bare surfaces were either conventional dry clean glass slides or polystyrene slide-sized pieces cut from disposable petri plates. Polymer films were made by immersing slides in filtered or centrifuged polymer solutions, allowing the slides to drain briefly and placing the slides flat in appropriate drying chambers. HA13 slides were dried in a chamber with anhydrous CaCl$_2$ for at least 12 hours. PHEMA and Nafion slides were dried in a dust free chamber and polyimide slide were dried in an oven at 100 °C. Before use, dry coated slides were soaked in a isotonic saline solution (0.9 wt.-% NaCl in water) to hydrate the polymer films.

Cell Suspension Preparation. Lymphocytes were isolated from the mesenteric lymphnodes of male Wistar rats (5 to 6 weeks old) by standard methods following anesthesia with chloral hydrate (400 mg/kg i.p.) and exsanguination through cardiac puncture. The collected lymphocytes were suspended in Hank's balanced salt solution (HBSS; Ca^{+2} and Mg^{+2}; pH 7.3) at a concentration of 1.5 ± 0.2 x 10^7 cell/mL. Cell viability was tested periodically by trypan blue exclusion and found to be at least 95%. All lymphocyte enumeration was done with manual cell counting with a hemocytometer.

L1210 murine leukemia wild type (L1210/0) and cisplatin (an important cancer chemotherapeutic agent) resistant (L1210/DDP) cell lines were maintained in culture through conventional techniques. The L1210 cell line is a commonly used in cancer research and is easily grown in culture as a cell suspension. Cells were harvested, washed twice with Dulbecco's phosphate buffered saline (PBS; pH 7.4; calcium and magnesium free) and suspended at a concentration of approximately 1.0 x 10^6 cell/mL. All L1210 cell enumeration was done with a Coulter Counter (Coulter Electronics, Hialeah, FL; Model ZF).

Separation Experiments. For each experiment a channel was constructed by sandwiching the Teflon spacer between the top plate and a never-used pre-soaked slide. Both during channel assembly and in subsequent flushing, great care was taken to ensure no air bubbles were trapped in the channel. After channel assembly, the channel was connected to the flow system and thoroughly flushed with HBSS for lymphocyte experiments or PBS for L1210 experiments. The complete flow system consisted of one or two constant flow pumps connected by Teflon tubing to a Teflon valve. The valve was used to either by-pass carrier flow around the channel or to switch to carrier streams of different flow rates. The valve was connected to the channel by a piece of oversized tubing which could be force fitted to the inlet tubing projecting from the channel. In separation experiments, the channel was by-passed from the carrier stream and disconnected from the flow system. The channel was then either filled completely with a lymphocyte suspension or with a fixed volume (40 μL) of an L1210 cell suspension. The channel was allowed to remain quiescent for a given amount of time (the stop-flow time). Immediately after the start of this period the channel was carefully reconnected to the flow system with the force-fitted tubing with care being taken to prevent disturbance of the channel or the inclusion of air bubbles in the inlet or outlet regions. At end of the stop-flow period the carrier flow was diverted through the channel and fractions were collected for periods ranging from 0.5 to 5 minutes. The volume of the fractions was determined gravimetrically and the cell concentration was determined by manual counting or Coulter counting. The majority of the experiments were conducted with the channel immersed in an ice bath before and during the experiment while other experiments were conducted at abient temperature.

Identification of B and T Cells. B cells were identified by immuno-fluorescent staining with fluorescein isothiocyanate (FITC) labeled rabbit anti-rat immuno-globulins. Determination of the percentage of fluorescent labeled cells in the initial cell suspension and in collected fractions was done by manual counting under a fluorescent microscope. Since the lymphocyte population in the mesenteric lymphnodes is almost entirely B and T cells with less than 5% null cells, all non-fluorescent cells were assigned as T cells (6).

Results

Cell Recovery from Non-HEMA Surfaces. Recovery of cells from non-HEMA (neither PHEMA or HA13) surfaces was poor. Lymphocytes and L1210 cells were only poorly recovered from glass (Figure 2). Recovery of L1210 cells (Figure 3) from polystyrene, polyimide coated glass, or Nafion coated glass, was only marginally better. This occurred despite the experimental conditions which are generally favorable for cell recovery including short stop-flow time (3.5 min), low temperature (0 °C, and high flow rate (> 5 cm/s).

Cell Recovery from HEMA Surfaces. In contrast to non-HEMA surfaces, both PHEMA and HA13 gave excellent cell recoveries. Indeed for L1210/0 and L1210/DDP, PHEMA was an almost non-adherent surface at 0 °C and short stop-flow times (3.5 min) when coated on either glass or polystyrene (Figure 4). Increasing the stop-flow time to 5 or 10 min did not greatly influence L1210/0 recovery from PHEMA coated glass (Figure 5) nor did increasing the temperature to 24 °C (data not shown).

These results are in marked contrast to those we have reported earlier for lymphocytes recovery from HA13 (Figure 6) (5). In these experiments, lymphocyte recovery was markedly diminished by both increasing temperature (4 °C to 23 °C) and increasing stop-flow time (3.5, 5, and 10 min).

Surface Selectivity and Cell Separation. HA13 shows substantial surface selectivity between B and T lymphocytes when use in column chromatography (4). This selectivity is also apparent in FFF/AC (Figure 7) as reported earlier (5). At a short stop-flow time (3.5 min; Figure 7a), B and T lymphocytes are easily detached while at long stop-flow times (10 min; Figure 7c) B cell are strongly adsorbed and T cells only poorly recovered. At an intermediate time (5 min; Figure 7b), both T and B cells can be completely recovered with T cells being recovered at a substantially lower flow rate than the B cells.

The pronounced flow rate lag between 100% T cell recovery and 100% B cell recovery for the case of a 5 minute stop-flow was exploited to allow separation of B and T cells. In the two separation experiments, an intermediate stop flow time (4.5 and 4.75 minutes; both at 4 °C) was used with an initial carrier flow of 1.0 cm/s. A 40 s fraction was collected and the flow rate was then stepped to a 4.6 cm/s and a 60 s fraction was collected. The first fraction in each case gave a T cell yield of 100% and a T cell purity of 99% while the second fraction gave a B cell yield of 70 and 99% respectively and a purity of 99%. The

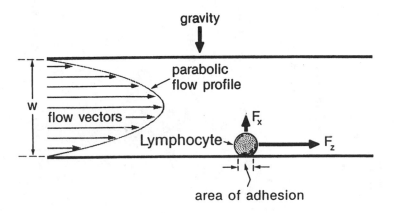

Figure 1. Schematic representation of flow and hydrodynamic forces in an FFF channel. (Reproduced with permission from reference 5. Copyright 1989 Elsevier Biomedical Division.)

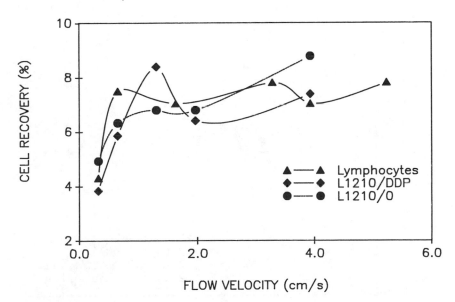

Figure 2. Lymphocyte and L1210 cell recovery versus flow rate from a glass accumulation wall surface. The stop-flow time was 3.5 min. and the temperature was 0 °C.

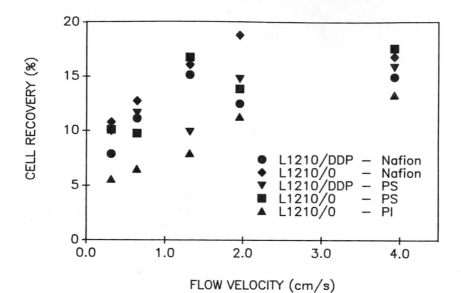

Figure 3. L1210 cell recovery versus flow rate from various accumulation wall surfaces at a stop-flow time of 3.5 min. at 0 °C. PS—bare polystyrene; Nafion—Nafion coated glass; PI—polyimide coated glass.

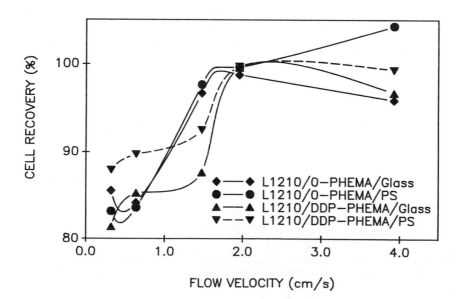

Figure 4. L1210 cell recovery versus flow rate from a PHEMA coated glass and PHEMA coated polystyrene (PS) accumulation wall surfaces at a stop-flow time of 3.5 min. at 0 °C.

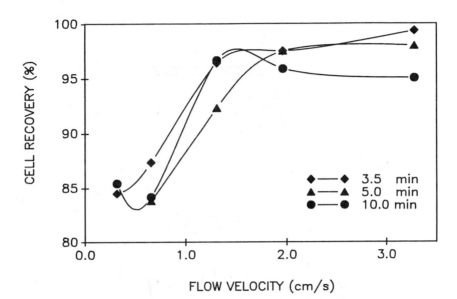

Figure 5. L1210/0 recovery versus flow rate from PHEMA coated glass accumulation wall surfaces at 0 °C at various times.

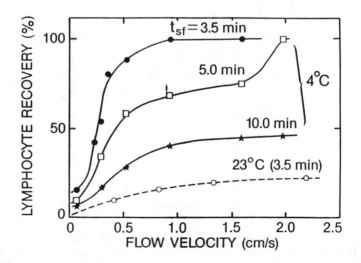

Figure 6. Total lymphocyte recovery from an HA13 coated glass accumulation wall as a function of flow rate for various conditions of stop-flow time and temperature. (Reproduced with permission from reference 5. Copyright 1989 Elsevier Biomedical Division.)

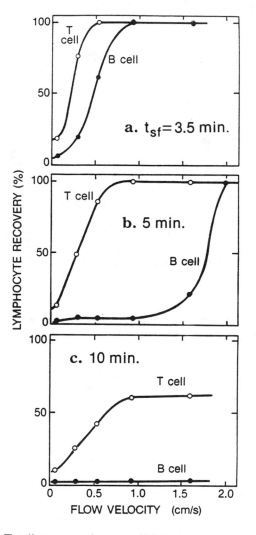

Figure 7. B and T cell recovery from an HA13 coated glass accumulation wall versus flow rate for various stop-flow times. The temperature was 4 °C. (Reproduced with permission from reference 5. Copyright 1989 Elsevier Biomedical Division.)

B cell fraction was however dilute and had debris. Approximately 1×10^6 cells were recovered in each experiment.

Discussion

The data in this report demonstrate that hybrid FFF/AC can be used in a number of ways. Those most obvious include the general evaluation of cell adhesion as demonstrated by the adhesive behavior of L1210 cells for various surfaces, and mammalian cell separation, as demonstrated by the separation of T and B cells. Applications which are less obvious, but perhaps equally important, include the use of FFF/AC as a means of quantitative measurement of adhesion forces and use of the technique as a general tool in membrane biology.

The results of cell-surface adhesion evaluations were surprising in several respects. While it was expected that there would be poor recovery from bare glass, the poor recovery from polystyrene and Nafion coated glass was unexpected. Polystyrene is commonly used in cell culture and in one of our laboratories (University of Vermont) L1210 cells are easily detached from polystyrene surfaces by gentle, though probably turbulent, swirling. Indeed, separation of L1210/0 and L1210/DDP was initially thought possible since L1210/DDP exhibit considerably more adhesion to polystyrene in culture than L1210/0 and show much more pronounced pseudopod formation on polystyrene surfaces. Nafion, a hydrophobic polymer with anionic sites, was expected to show substantial non-adherence since both the polymer and cells carry net negative charges on their surface. The result from FFF/AC indicate these views of cell adhesion are simplistic.

As a cell separation method, hybrid FFF/AC clearly benefits from the use of FFF-type channels for AC separations. The unpacked nature of these channels allows flow rates to be easily and rapidly increased in a step-wise fashion. This step-wise elution program yielded a very good separation of B and T cells. The relatively large flow-step required (1.0 to 4.6 cm/s) is indicative of the slow rise of the recovery/flow-rate curves. Much of the broadness in rise time is probably the result of the fairly rapid kinetics of cell adhesion. While this kinetic induced broadening was minimized by performing experiments at low temperature, there is still a 2 to 3 min time differential between cells that must sediment the entire thickness of channel before undergoing the adhesion process and cells that reach the channel surface almost immediately on injection. This problem can perhaps be mitigated by the use of a physical split injection system as is employed in several instrument in the laboratory of one of our groups (University of Utah) (7).

The apparent rapid kinetics observed with lymphocyte adhesion to HA13 as well as the rapid adhesion observed with all cells on non-HEMA surfaces indicate FFF/AC may also used as a tool to study rapid biological kinetics of cell surface adhesion, a largely unexamined area. Since cell adhesion is critical in many areas including cancer metastasis and thrombosis research, a tool for the study of adhesion kinetics could be of considerable value.

A limitation of the lymphocyte separation is the relatively small yield of cells for each lymphocyte experiment, approximately 1 million cells versus about 10 million cells for the column chromatography method. However, the channel used in these experiments are much smaller than those normally used in FFF and were selected primarily to be easily and rapidly constructed to allow new accumulation walls to be used for each experiment. Conventional FFF channels have a much greater volume (typically 2 mL to 4 mL) and could give cell yields as great as that found with the column method. Additionally, larger FFF channels would allow the dispersive-force/driving-force separating ability of FFF to be applied to the cell separation process, a powerful aspect of FFF not employed in this study. If kinetic broadening can be minimized, it is likely that flow rate programming can be used to induce controlled particle detachment and give high resolution FFF/AC separations. The ability of FFF/AC to screen cell/surface combination could also be utilized to determine optimal surfaces for pure FFF separations. For instance, the L1210 adhesion data make it clear that PHEMA coated glasses would be a better choice for FFF separations of cells than bare glass. The results also indicate that polyimide, frequently used as a tape to cover the accumulation walls of sedimentation FFF channels, may not be a good choice for FFF cell separations.

The use of FFF-type channels allows the hydrodynamic shear stress to be calculated at any point in the channel by the equation

$$\tau = 6<v>n/w \tag{1}$$

where w is the channel thickness, $<v>$ is the mean cross sectional flow velocity, and n is viscosity. Goldman et al. (8) and O'Neill et al (9) have derived an expression for the viscous-drag force on a particle of radius a acting to pull the particle in the direction of the flow.

$$F_z = 32.06a^2\tau \tag{2}$$

The torque acting on the particle is given by Hubbe (10) as

$$\tag{3}$$
$$T = 43.92a^3\tau$$

Saffman (11) has also derived an expression for the lift force (4).

$$F_x = 6.46a^3n^{-1}p^{\frac{1}{2}}\tau^{3/2} \tag{4}$$

It should be noted that this equation is strictly valid only for flow occurring far from a wall although Hubbe justifies its use for adhered particles. Using the value of the flow rate at the estimated inflection points of the T cell (0.36 cm/s) and B cell (1.8 cm/s) recovery curves in Figure 7b, one can calculate the shear stress near the walls of the channel for these flow rates. Taking the radius of an average lymphocyte to be 4 μm and using tabulated values, equations 1-3 can be solved. The results of these calculations are presented in Table I. It is of note that τ is linearly dependent on flow velocity and that both F_z and torque are linearly dependent on τ while F_x is dependent on τ to the 3/2 power. Thus the equations 1 and 2 predict that the B cells are bound 5 times more strongly than T cells while equation 3 predicts that B cell are bound approximately 11 times more strongly than T cells.

It can be seen from Table I that the value of T is one to two orders of magnitude smaller than that of either F_x or F_z while F_x is approximately 10 times less than F_z. Thus, it would appear from this that the cell binding force is dominated by F_z and the other forces may be neglected.

While there is no apriori reason to suspect the values of the binding forces in Table I, they must nevertheless be treated with a caution. It is possible the equations from which these forces were calculated are not valid under the conditions of these experiments or that unaccounted for factors such as micro-scale surface roughness or particle deformation seriously compromise the validity of these calculations. There is currently no independent means of verification of these calculated forces and without such a means these values should not be greatly relied upon as true force values. However, if the general form of these equation are still valid, a much less stringent requirement, then the relative force differences can be taken with more confidence. With this in mind it is a reasonable conclusion than B cells, under the experimental conditions used here, are adhered 5 to 10 times more strongly on the HA13 surface than T cells.

Table I. Values of critical flow velocity and force parameters corresponding to 50% cell

Parameter	T Cell	B Cell	T Cell/B Cell Ratio
Flow Velocity	0.36 cm/s	1.8 cm/s	5
F_z	6.9×10^{-6}	3.4×10^{-5}	5
T	3.8×10^{-9}	1.8×10^{-8}	5
F_x	4.1×10^{-8}	4.4×10^{-7}	11
Shear Stress	1.3	6.6	5

Force is in dynes and torques in dyne-cms. Shear stress is in dyne cm^{-2}.
Results for 5 min stop-flow time; HA13-coated slide glass at 4 °C. Adapted from ref. 5.

Acknowledgements

The work at the University of Utah was supported by the National Institutes of Health Grant GM10851-31. The work at the University of Vermont was supported by the National Cancer Institute Vermont Regional Cancer Center Core Grant PHS CA-22435-09.

Literature Cited

1. Giddings, J.C.; Chen, X.; Wahlund, K.-G.; Myers, M.N. *Anal. Chem.* **1987,** *59,* 1957.
2. Kataoka, K.; Okano, T.; Sakurai, Y.; Nishimura, T.; Maeda, M.; Inoue, S.;Watanabe, T.; Tsuruta, T. *Makromo Chem.* **1982,** *3,* 275.
3. Kataoka, K.; Okano, T.; Sakurai, Y.; Nishimura, T.; Inoue, S.;Watanabe, T.; Maruyama, M.; Tsuruta, T. *Eur. Poly. J.* **1983,** *19,* 979.
4. Maruyama, A.; Tsuruta T.; Kataoka, K.; Sakurai, Y. *Makromol Chem.* **1987,** *8,* 27.
5. Bigelow J.C.; Giddings C.G.; Nabeshima, Y.; Tsuruta, T.; Kataoka, K.; Okano,; Nobuhiko, Y.; Sakurai, Y. *J. Immunol. Methods.* **1989,** *117,* 289.
6. Hudson, L.; Hay, F.C. *Practical Immunology;* Blackwell: Oxford, 1976.
7. Giddings, J.C. *Sep. Sci. Technol.* **1985,** *20,* 749.
8 Goldman, A.J.; Cox, R.G.; Brenner, H. *Chem Eng. Sci.* **1967,** *22,* 653.
9. O'Neill, M.E. *Chem. Eng. Sci.* **1968,** *23,* 1293.
10. Hubbe, M.A. *Colloids Surf.* **1984,** *12,* 151.
11. Saffman, P.G. *J. Fluid Mech.* **1965,** *22,* 385.

RECEIVED March 15, 1991

Chapter 11

High-Capacity Cell Separation by Affinity Selection on Synthetic Solid-Phase Matrices

Kazunori Kataoka

Department of Materials Science and Research Institute for Biosciences, Science University of Tokyo, Noda, Chiba 278, Japan

There has been a growing demand in the field of biomedical sciences to prepare highly pure and viable cell populations. An important facet of this study is to develop new polymeric adsorbents with specific affinity toward a particular subpopulation of lymphocytes. Based on our strategy of separating lymphocyte subpopulations through their differential ionic affinity toward solid-phase matrices, a series of poly(2-hydroxyethyl methacrylate)/ polyamine graft copolymers (HA copolymers) was prepared. HA copolymer columns were found to show specific adsorption affinity toward B cells from a mixture of B and T cells, and allows for separation of B and T cells in high yield and purity with a short operating time. In addition, a partially quaternized HA copolymer (HQA) was found to have a capability of resolving T cell subsets (helper and suppressor T cells). These results demonstrate the promising features of cell adsorption chromatography as a major tool for separating a wide spectrum of cell populations.

With recent progress in the biomedical sciences, the development of cell separation systems becomes especially important because of the essential role of pure and viable cell populations for assay and therapy of various diseases as well as for production of high-value bioactive compounds(1-3). Among the great many cell populations of interest, lymphocyte subpopulations are the ones whose separation has received the most practical interest because their successful separation is the critical process in precise assays and effective therapy of serious diseases, especially those related to immune responses. For the treatment of autoimmune disorders, therapy called "lymphocytapheresis", removal of lymphocytes harmful to patients from their blood, has been performed increasingly(4). Recently, "adoptive immunotherapy", a new concept in cancer immunotherapy, was introduced by Rosenberg et al(5), in which peripheral blood lymphocytes of a patient were activated in vitro by interleukin-2 and reinjected into the patient.   The collection and concentration of lymphocytes from patients' blood is the key process in this novel immunotherapy.

In the field of diagnosis, lymphocyte subpopulations have been separated routinely for assays of cellular immunity as well as for human leukocyte antigen(HLA) typing in transplantation(6). Although nylon fiber columns have been utilized widely for the diagnostic separation of lymphocyte subpopulations, there still exist many problems related to yield, purity, and function of separated cells(7-10).

From the industrial point of view, lymphocytes are quite interesting because of their availability to produce high-value bioactive compounds including monoclonal antibodies, interferons, and interleukins. Thus, the preparative separation of lymphoid cells is one of the major processes in cellular biotechnology.

Important goals of our research include the development of new synthetic polymers with specific affinity toward a particular subpopulation of lymphocytes, and to apply these polymers as an adsorbent used for separation of lymphocyte subpopulations. Our strategy is to resolve lymphocyte subpopulations through their differential ionic interactions with polymers having basic amino groups.

Lymphocytes consist of many subpopulations with distinct immunological functions. B cells differentiate into antibody-forming cells, and are utilized in the biotechnology field for the preparation of hybridoma cell lines that produce monoclonal antibodies. Also, B cell enrichment is critical for the typing of HLA-DRw antigens. T cells are subdivided into many subclasses, including helper T cells, suppressor T cells, cytotoxic T cells, and DTH(delayed-type hypersensitivity) effector T cells. All of these subsets play an indispensable role in immune responses through activating or suppressing cellular immunity. Further, T cells have received considerable interest because they can be utilized for large-scale production of diverse types of lymphokines.

Although these subpopulations are different in their functions, they are similar in their shape, volume, and density. Thus, centrifugal techniques, the most popular for cell separation, are not effective for their separation, and for this reason, a more effective separation technique for lymphocyte subpopulations has been strongly desired. Worthy of mention is that even morphologically indistinguishable cell populations often have distinctive surface properties due to differences in their plasma membrane composition. This may lead to differential cellular adsorption onto materials with distinctive chemical structures. We gave special attention to the differential ionic character of B and T cell surfaces arising from differences in the number and species of acidic functional groups expressed on their plasma membranes(11). Thus, basic amino groups were chosen as the main component for the design of polymeric materials expected to have specific ionic interactions with lymphocyte subpopulations.

Microdomain-structured Polymers as Adsorbents for Cell Separation

The interaction of cells with material surfaces consists of two distinct successive stages(1). The first stage is an adsorption process in which the interaction is mediated through physicochemical forces (stage I).   The subsequent second stage is a metabolism-dependent process, which we define as "adhesion" (stage II). Adhesion (stage II) involves rearrangement and reorganization of cytoskeletal components, accompanied by changes in cellular functions as well as in cellular shape, leading to a markedly spreading morphology.   These sequential processes are termed "contact-induced activation" because they are triggered by cellular contact with material surfaces. From the standpoint of cell separations based on a reversible adsorption /desorption procedure on an adsorbent, this contact-induced activation is unfavorable because it is likely to increase non-specific cellular adhesion, to decrease cellular recovery, and to damage functions of separated cells.   Thus, in addition to high specificity, cellular adsorbents need to have abilities to inhibit or at least retard adsorbing cells from proceeding into stage II or contact-induced activation.

Materials that do not cause undesirable cellular activation have been studied in our research group in the course of developing antithrombogenic materials(12,13). For this purpose, platelets, a central cellular component in thrombosis, should be strictly prevented from proceeding into contact-induced activation. Recently, we have shown that microdomain-structured polymers made by block or graft copolymerization techniques have an ability to prevent platelets from proceeding into the contact-induced activation state(12,13), leading these polymers to have excellent antithrombogenicity.

As a possible mechanism involved in the suppressive effect of microdomain-structured surfaces on cellular activation, we have proposed a mechanism of "capping control"(1). Cellular plasma membrane has a microheterogeneous structure composed of a continuous phase of lipids with dispersed glycoprotein molecules, most of which are able to move laterally through the lipid moieties(14).   The change in the lateral assemblage of these membrane proteins is considered to play an essential role in the transfer of stimuli through the membrane to the cellular interior, leading the change in cellular metabolism. Based on this close relationship between cellular metabolism and assemblage of membrane proteins, we proposed a mechanism in which microdomain structures on polymer surfaces may restrict the change in cellular energy metabolism by regulating excessive assembly (cap formation) of membrane proteins at the cell/materials interface(1).

This unique feature of microdomain-structured surfaces preventing cells from contact-induced activation leads us to the idea of "functional microdomains", where cellular specific functional groups are fixed on one of the microdomains as schematically shown in Fig. 1.   Specific adsorption as well as least undesirable activation of particular cell populations would be achieved by using adsorbents with this type of "functional microdomains". We focused on amino groups as the functional moiety to interact with lymphocyte subpopulations through acidic moieties on the lymphocyte membrane surfaces.   Thus, polymers having amino-derivatized microdomains were prepared for this purpose.

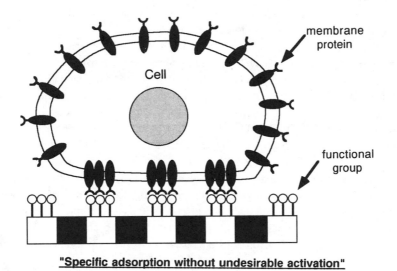

Fig. 1 Schematic illustration of "functional microdomains" and their interaction with living cells.

Polyamine Graft Copolymers as Novel Polymeric Adsorbent for Separation of Lymphocyte Subpopulation

Graft copolymers giving amino-functionalized microdomains in their film form were prepared by radical copolymerization of vinyl-terminated polyamine (polyamine macromonomer) **1** with corresponding comonomer, styrene **2** or 2-hydroxyethyl methacrylate(HEMA) **3** (Fig. 2)(15). The polyamine composition in the copolymer is readily controlled by changing the feed ratio of macromonomer to the corresponding comonomer during polymerization. Further, we can control the chain length of the polyamine graft in the copolymer by utilizing polyamine macromonomers with varying molecular weights(16). An elegant synthesis of this polyamine macromonomer was established by Nitadori and Tsuruta based on the lithium amide-catalyzed polyaddition reaction of p-divinylbenzene and N,N'-diethylethylenediamine(17).   Polyamine macromonomers with the desired molecular weight were prepared by adjusting the ratio of [>NH] to [>NLi] in the reaction mixture.   A detailed review by Tsuruta on the reaction mechanisms is available(18). Consequently, polyamine graft copolymers with controlled composition including the content and the chain length of the polyamine graft were prepared and used in the study on lymphocyte separation.   Unless otherwise noted, copolymers having a polyamine graft with 5000 mol wt have been utilized for the following studies.

As reported elsewhere(15), film surfaces of these graft copolymers have a sea-and-island type of microdomain structure: islandous-shaped polyamine microdomains with diameters ranging in the order of ca. $10^{-1}\mu m$ are dispersed in a continuous "sea-like" phase of polystyrene or PHEMA. The polyamine islands are considerably smaller than lymphocytes, whose diameter is approximately 7 μm. Thus, a lymphocyte coming into contact with graft copolymer surface may have multi-sited interactions with polyamine microdomains as schematically shown in Fig. 1.

Adsorbents for the separation of lymphocyte subpopulations were prepared by coating glass beads of 48 to 60 mesh with polyamine graft copolymer. Adsorbents thus prepared were packed into polyvinylchloride tubing (inner diameter: 3 mm) to make columns utilized for lymphocyte separation. Lymphocytes prepared from rat mesenteric lymphnodes were mainly used in this study.

We focused on the differential retention of two major subpopulations of lymphocytes, B cells and T cells. In the course of our study, we have found the following relationship to hold(19):

$$-\log([B]/[B]_0) = -(A_B/A_T)\log([T]/[T]_0) \tag{1}$$

where A_B/A_T is the selectivity index, $[B]/[B]_0$ and $[T]/[T]_0$ are the effusion of B and T cells from the column, respectively. Obviously, the value of A_B/A_T increases with increased affinity of the adsorbent toward B cells. The value of A_B/A_T is determined as

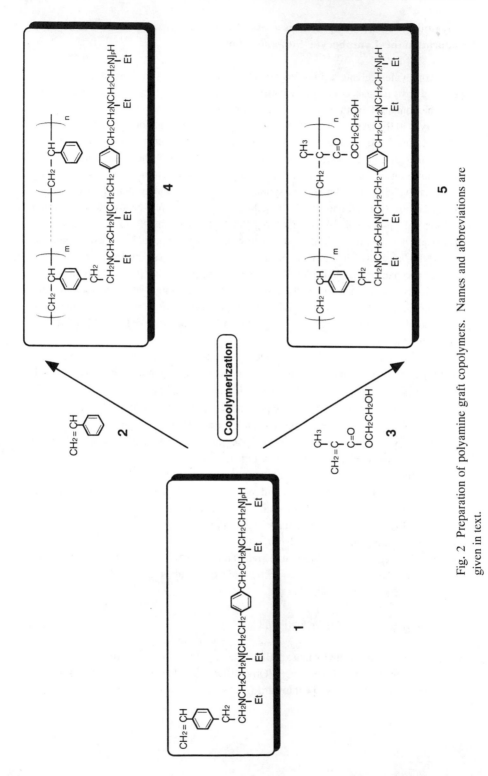

Fig. 2 Preparation of polyamine graft copolymers. Names and abbreviations are given in text.

the slope of the straight line obtained by plotting -log($[B]/[B]_0$) against -log($[T]/[T]_0$). The results of this plot for albumin-coated columns of polystyrene(PSt) and polystyrene/polyamine graft copolymer **4** with polyamine content of 50 wt% (SA 50), respectively, are shown in Fig. 3.   An SA 50 column has a selectivity index of approximately 4, indicating its high selectivity toward B cells.   However, this excellent selectivity of the SA copolymer column appears only when the column is coated with albumin prior to lymphocyte loading. Without albumin treatment, the SA copolymer column has a selectivity index as low as 1.5, the same value as a PSt column.   This result indicates the influence of the hydrophobicity of the matrix polymer.   Hydrophobic interaction of lymphocytes with PSt domains on the SA surface is considered to be responsible for non-specific retention of T cells as well as B cells in SA copolymer columns without albumin treatment, resulting in the low selectivity. Albumin may mask the hydrophobic interaction through its adsorption on the PSt domains of the SA copolymer surface, allowing ionic interaction to become the main driving force of lymphocyte adsorption.   Thus, a more hydrophilic backbone polymer than PSt is favorable for the resolution of B and T cells based on their differential ionic affinity toward polyamine moieties.   Consequently, we then prepared a graft copolymer with poly(2-hydroxyethyl methacrylate) (PHEMA) as the backbone polymer.   The structural formula of PHEMA/polyamine graft copolymer (HA copolymer) **5** is shown in Fig. 2.   As expected, the HA column has a high affinity toward B cells even without albumin treatment(20). The separation profile was examined by flow cytometry, where B cells show higher fluorescence intensity due to the binding of   fluorescein isothiocyanate(FITC)-tagged antibody toward rat immunoglobulin molecules expressed on the membrane surface of B cells.   The peak corresponding to B cells   virtually disappeared in the column effluent, demonstrating the selective retention of B cells on the HA copolymer column. As summarized in Table I, a T cell population with more than 95% purity was obtained as effluent in a yield of nearly 100%.   Worthy of further mention is the simple operational condition of the HA-column system.   Passage of lymphocytes through the column takes less than 5 min, ensuring a low incidence of cellular damage due to a prolonged handling process.   Both SA and HA columns worked successfully even at temperatures as low as 4°C.   This result strongly indicates that the resolution of B and T cells by these graft copolymer columns is based on the differential physicochemical affinity of B and T cells on graft copolymer surfaces and not on the differential energy-dependent activation of these cells triggered by the contact with these surfaces(21).

Probably due to the "microdomain-effect", lymphocytes attached on graft copolymer surfaces maintain their original spherical shape,   in   contrast to the remarkable shape changes, including flattening and pseudopod formation, of lymphocytes on a PHEMA surface. This feature offers the great advantage of high recovery of attached B cells from the column (22). Indeed, as can be seen from Table 1, a B cell population with more than 90% purity and 90% yield was successfully recovered from an HA copolymer column by a gentle pipetting procedure. The "microdomain-effect" of graft

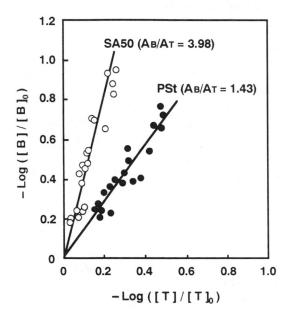

Fig. 3 -Log ([B]/[B]$_0$) vs -log([T]/[T]$_0$) plots for PSt column (closed circles) and
SA 50 column (open circles) with albumin coating

Table I. Purity and Recovery of B and T cells separated by HA Copolymer Column

pH of cell suspension	T cells		B cells[a]	
	purity (%)	recovery (%)	purity (%)	recovery(%)
7.0	92.4±3.1	69.4±3.2	60.6±3.8	94.7±4.0
7.2	93.1±3.5	95.7±2.9	90.1±3.7	93.1±2.1
7.3	96.3±2.7	99.2±0.8	97.3±1.0	92.4±2.3

a) B cells are recovered from the column by gentle pipetting with addition of 1% bovine
serum albumin

copolymers, in which cellular activation is remarkably blocked, was further confirmed through detailed studies done to clarify the effect of cellular metabolism on their interaction with polymer surfaces(21, 23).

Recently, the HA copolymer column was found to be useful for the separation of murine lymphocytes as well as rat lymphocytes (manuscript in preparation). Further, it has been utilized for the collection of natural-killer (NK) cells from human peripheral blood(24).

Separation Mechanisms of Polyamine Graft Copolymer Columns

It is important to gain insight into the mechanisms involved in the separation of B and T cells by graft copolymer columns. Because lymphocyte adsorption on these copolymers was significantly affected by pH and the ionic strength of the medium, ionic interaction is suggested to have a principal role. Most likely is the ionic interaction of protonated amino groups in the copolymer with acidic groups on the cellular plasma membrane surface. Thus, the protonation status of amino groups in the graft copolymer was elucidated by acid-base titration of polyamine macromonomer 1 (25), which has an ethylene diamine moiety in each repeating unit. Consequently, corresponding to the two-stage protonation of the amino groups in the ethylene diamine moiety, a two-step titration curve with plateaus at pH values around 5 and 7, respectively, was obtained. Degree of the protonation, $\alpha(=[\text{protonated amino groups}]/[\text{total amino groups}])$, at various pH values was determined from this titration curve (Fig. 4). The α-pH curve in Fig. 4 indicates that the polyamine chain considerably changes its degree of protonation in the physiological pH range, where most cell separation experiments were carried out. The half equivalence point ($\alpha=0.5$) of the polyamine macromonomer was found to be in the pH range of 6.4 to 7.0, depending on the molecular weight; shifts were to lower pH with increasing molecular weight (16) (see Fig. 5).   Further, the half equivalence point virtually coincides with the phase transition point (turbid point) of polyamine chains in the aqueous environment.   That is, across this point ($\alpha=0.5$), the polyamine changes its conformation from the water-soluble extended state (at $\alpha>0.5$) to the water-insoluble aggregated state (at $\alpha<0.5$). This correlation is valid for polyamine samples with varying molecular weight as shown in Fig. 5. This unique feature of polyamine macromonomer 1 is indeed reflected in the interfacial characteristics of graft copolymers containing polyamines as grafted segments. Fig. 6 illustrates possible shapes of the polyamine chain in high and low protonation status, respectively, at the aqueous interface of the graft copolymer.

This conformational change of the polyamine chain has a striking effect on the retention of lymphocytes on the HA copolymer column, as shown in Fig. 7 (25). The sudden increase in T cell retention in the range of $\alpha>0.5$ is well correlated with the conformational change of the polyamine chain from the aggregated state to the extended state. That is, the T cell interacts negligibly with polyamine in the aggregated state. This is in sharp contrast with the B cell which interacts considerably with polyamine in

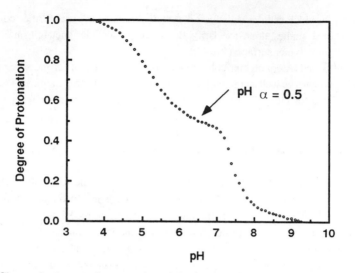

Fig. 4 Change in degree of protonation (α) of polyamine macromonomer **1** with pH

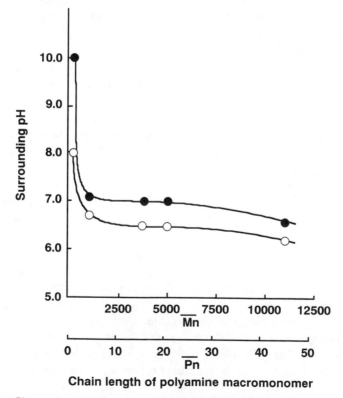

Fig. 5   Change in turbidity point (closed circles) and half equivalence point (pHα=0.5) (open circles) with chain length of polyamine macromonomer **1**

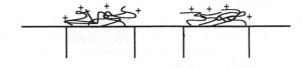

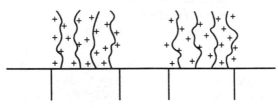

Fig. 6 Schematic morphology of polyamine chains in functional microdomains at aqueous interface of graft copolymer

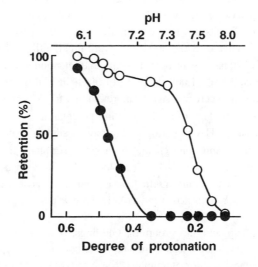

Fig. 7 Effect of degree of protonation (α) on retention of B cells (open circles) and T cells (closed circles) on HA13 column

the aggregated state as well as in the extended state. Consequently, in the range of $\alpha=0.3\text{-}0.4$, which approximately corresponds to the physiological pH range of 7.0-7.4, the HA column adsorbed more than 90% of loaded B cells with negligible retention of T cells, resulting in a T cell-enriched population with more than 95% purity in nearly 100% yield as the column effluent.

A comparative study using random copolymer **6** (Fig. 8) of HEMA and N,N'-diethyl-N-(4-vinylphenethyl)ethylenediamine (DEVED) as the cellular adsorbent further verifies the crucial role of the pH-induced conformational change of the polyamine chain in the high resolution separation of B and T cells (16). Fig. 9 shows the retention of B and T cells on random copolymer **6** with different amounts of DEVED and measured under varying pH conditions. Data are normalized for the content of protonated nitrogen (N^+) in the copolymer calculated by multiplying the N content in copolymer **6** by α at the given pH. Although B cells always showed slightly higher retention than T cells on copolymer **6** with varying N^+ content, the difference is considerably smaller than that observed for the HA copolymer column (see Fig. 7) and, from the view point of separation, the purity of the cell population was inadequate for practical usage. Consequently, these results clearly demonstrate that pH-induced conformational change of the polyamine chain as well as the protonation degree of amino groups are the crucial factors for the high resolution of lymphocyte subpopulations by HA copolymer columns.

The molecular weight or chain length of the polyamine graft has a remarkable effect on B and T cell retention (16). We have recently determined that B cell retention increases with an increase in the molecular weight of polyamine graft from 890 to 11,000. Retention of T cells has no such simple correlation with molecular weight (M.W.) In the M.W. range of 890-6,600, there were no significant differences in the profile of T cell retention; namely, there was negligible retention at $\alpha<0.5$. However, HA 13 with polyamine M.W. 11,000 [HA 13 (11,000)] had different behavior with respect to T cell retention. There was significant retention of T cells on HA 13 (11,000) even in the range of $\alpha<0.5$. The result of streaming potential measurements suggested that the polyamine chain of HA 13 (11,000) may have a slightly expanded state even $\alpha<0.5$ due to its considerably long chain length (manuscript in preparation). The unexpected high T cell retention on HA 13 (11,000) might be due to this anomalous conformational state of the polyamine graft. Consequently, the most efficient separation of B and T cells was accomplished using HA 13 with a polyamine M.W. range of 3,000-6,600.

Recently, our group and the group of Giddings, University of Utah, have collaborated to develop a new cell separation method named hybrid field-flow fractionation (FFF)/adhesion chromatography (26). A thin ribbon-like chamber with an open-flow channel was used in this method. The inner-bottom surface of the chamber was coated with HA 13. Lymphocytes were introduced into the chamber, settled for a given time period and subsequently recovered from the chamber by stepwise increases in flow. By regulating the mode of stepwise flow increase, complete separation of B and T cells was achieved by this new method. Further, the attachment strength of B and T cells on the HA surface could be evaluated by this method, revealing that B cells adsorb about 5 times as strongly on HA surface as T cells. The details of the hybrid FFF/adhesion chromatography are described in another chapter of this book.

Column Separation of T Cell Subsets using A Partially Quaternized HA Copolymer (HQA) as Adsorbent

Our most recent interest focused on the design of synthetic adsorbents for T cell subsets, including helper, suppressor, and killer T cells. There is a great demand in biomedical sciences to isolate viable T cell subsets with high yield and purity for their functional assay as well as for clinical usage including diagnosis and therapy of immuno-diseases.   Although, the monoclonal antibody-based method for separating lymphocyte subsets has become a powerful tool, there still exist many disadvantages including unfavorable biological responses, low storage stability, and high cost for preparative process. These disadvantages can be alleviated through the development of synthetic cellular adsorbents with an adequate affinity toward a particular T cell subsets. Recently, we have found that the partially quaternized HA copolymer (HQA) **7** (Fig. 10) can be used for resolving helper and suppressor T cells (manuscript in preparation).

In contrast to HA, there always exists a certain amount of adsorbing T cells on HQA even in the physiological pH range, where the cationic fraction,

$$F^+ = ([>N^+H-] + [-N^+R-])/([>N-] + [>N^+H-] + [>N^+R-]) \qquad (2)$$

of amino groups in the polyamine chain is below 0.5. Aqueous solutions of partially quaternized polyamine show no definite turbidity in the high pH region, suggesting that it may still have considerably extended conformation even when F^+ is below 0.5. This is in sharp contrast with non-quaternized polyamine which shows a clear conformational change with α. It is plausible that partially quaternized polyamine chains may considerably extend into the aqueous interior at the HQA/water interface even at physiological pH, resulting in a certain interaction with T cells. Irrespective of the physical process, this interaction is T cell-subset dependent.

Immuno-fluorescence assay using FITC-labeled monoclonal antibody revealed that at pH 7.4 HQA preferentially adsorbed suppressor T cells as well as B cells (derived from the lymphnode of a Wistar rat). No such resolution was observed with the HA column, indicating a crucial role of the quaternary ammonium groups for this resolution. The degree of quaternization of amino groups also has an important effect. The sample with the highest degree of quaternization, 28.9% [HQA 13 (28.9)] yielded the most enriched helper T cell population with more than 60% purity as column effluent. Further, resolution of T cell subsets by HQA was considerably pH-dependent, and at a slightly acidic condition (pH=6.8) helper T cells in turn showed higher affinity to HQA

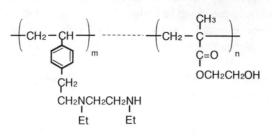

Fig. 8   Copolymer **6** of HEMA and N,N'-diethyl-N-(4-vinylphenethyl)-ethylenediamine

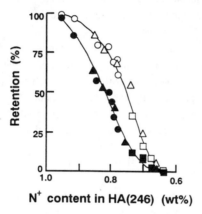

Fig. 9   B and T cell retention on random copolymer **6** with varying N^+content. Open symbols for B cells and closed symbols for T cells. O, ●: copolymer composed of 7 wt% DEVED - 93 wt% HEMA; △, ▲: 10 wt% DEVED-90 wt% HEMA; □, ■: 13 wt% DEVED-87 wt% HEMA

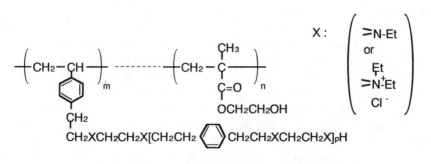

Fig. 10 Partially quaternized HA copolymer (HQA)

than suppressor T cells.   This result   indicates the possibility of pH-dependent resolution of T cell subsets.

Conclusions

In conclusion, our results demonstrated a successful separation of lymphocyte subpopulations through differential ionic affinity to synthetic adsorbents, indicating a promising feature of cellular adsorption chromatography as a major tool for separating a wide spectrum of cell populations.

Acknowledgement

Our work cited in this paper was performed in collaboration with the research groups of Prof. Teiji Tsuruta, Science University of Tokyo, and Prof. Yasuhisa Sakurai, Tokyo Women's Medical College.

Literature Cited

1. Kataoka, K.; Sakurai, Y.; Tsuruta, T. *Makromol. Chem., Suppl.,* **1985,** *9,* 53
2. Kataoka, K.; Sakurai, Y.; Tsuruta, T. In *Proteins at Interfaces; Physicochemical and Biochemical Studies;* Brash, J.L. and Horbett, T.A., Eds.; ACS Symposium Series No. 343; American Chemical Society; Washington, D.C., 1987; p.604
3. Kataoka, K.; *CRC Critical Reviews in Biocompatibility,* **1988,** *4,* 341
4. Tindall, R.S.A., Ed.; *Therapentic Apheresis and Plasma Perfusion;* Alan R. Liss, New York, 1982
5. Rosenberg, S.A.; Lotze, M.T.; Muul, L.M.; Leitman, S.; Chang, A.E.; Ettinghausen, S.E.; Matory, Y.L.; Skibber, J.M.; Shiloni, E.; Vetto, J.T.; Seipp. C.A.; Simpson, C.; Reichert, C.M.; *N. Engl. J. Med.,* **1985,** *313,* 1485
6. Sabato, G.D., Langone, J.J.; Vunakis, H., Eds.; *Separation and Characterization of Lymphoid Cells;* Methods in Enzymology; Academic Press; Orlando, Fla., 1984; Vol. 108
7. Handwerger, B.S.; Schwartz, R.H. *Transplantation,* **1974,** *18,* 544
8. Indiveri, F.; Huddlestone, J.; Pellegrino, M.A.; Ferrone, S.,   *J. Immunol. Methods,* **1980,** *34,* 107
9. Corrigan, A.; O'Kennedy, R.; Smyth, H., *ibid.,* **1979,** *31,* 177
10. Hunt, S.V., In *Handbook of Experimental Immunology, 3rd ed.;* Weir, D.M., Ed.; Blackwell Scientific, Oxford, **1978,** Vol. 3; Chap. 24
11. Marchalonis, J.J.; Wang, A-C.; Galbraith, R.M.; Barker, W.C., In *The Lymphocyte, Structure and Function, 2nd ed.*; Marchalonis, J.J., Ed.; Marcel Dekker, New York, **1988,** p. 307

12. Yui, N.; Kataoka, K.; Sakurai, Y., In *Artificial Heart I*, Akutsu, T.; Koyanagi, H.; Pennington, D.G.; Poirier, V.L.; Takatani, S.; Kataoka, K. Eds.; Springer, Tokyo, 1986; p. 23

13. Yui, N.; Kataoka, K.; Sakurai, Y., *Med. Prog. through Tech.*, **1987**, *12*, 221

14. Singer, S.J.; Nicolson, G.L., *Science*, **1972**, *175*, 720

15. Maruyama, A.; Senda, E.; Tsuruta, T.; Kataoka, K.,*Makromol. Chem.*, **1986**, *187*, 1895

16. Nabeshima, Y.; Tsuruta, T.; Kataoka, K.; Sakurai, Y., *J. Biomater. Sci., Polym. Ed.*, **1989**, *1*, 85

17. Nitadori, Y.; Tsuruta, T., *Makromol. Chem.*, **1979**, *180*, 1877

18. Tsuruta, T., In *Polymeric Amines and Ammonium Salts;* Goethals, E.J., Ed.; Pergamon Press, Oxford, 1980; p.163

19. Kataoka, K.; Okano, T.; Sakurai, Y.; Nishimura, T.; Inoue, S.; Watanabe, T.; Maruyama, A.; Tsuruta, T., *Eur. Polym. J.*, **1983**, *19*, 979

20. Maruyama, A.; Tsuruta, T.; Kataoka, K.; Sakurai, Y., *J. Biomed. Mater. Res.*, **1988**, *22*, 555

21. Maruyama, A.; Tsuruta, T.; Kataoka, K.; Sakurai, Y., *Biomaterials*, **1989**, *10*, 291

22. Maruyama, A.; Tsuruta, T.; Kataoka, K.; Sakurai, Y., *Biomaterials*, **1989**, *10*, 393

23. Maruyama, A.; Tsuruta, T.; Kataoka, K.; Sakurai, Y., *Biomaterials*, **1988**, *9*, 471

24. Komiyama, K.; Oda, Y.; Iwase, T.; Yoshida, H.; Moro, S., *Igaku-no-Ayumi(in Japanese)*, **1988**, *146*, 809

25. Maruyama, A.; Tsuruta, T.; Kataoka, K.; Sakurai, Y., *Makromol. Chem., Rapid Commun.*, **1987**, *8*, 27

26. Bigelow, J.C.; Giddings, J.C.; Nabeshima, Y.; Tsuruta, T.; Kataoka, K.; Okano, T.; Yui, N.; Sakurai, Y., *J. Immunol. Methods*, **1989**, *117*, 2289

RECEIVED March 15, 1991

Chapter 12

Factors in Cell Separation by Partitioning in Two-Polymer Aqueous-Phase Systems

Derek Fisher[1,2], F. D. Raymond[1,4], and H. Walter[3]

[1]Department of Biochemistry and [2]Molccular Cell Pathology Laboratory, Royal Free Hospital School of Medicine, University of London, Hampstead, London NW3 2PF, England
[3]Laboratory of Chemical Biology, Veterans Affairs Medical Center, Long Beach, CA 90822

The partitioning of cells in two-polymer aqueous phase systems depends on cell-phase specific interactions and their effect on the kinetics of phase separation in the early stages following mixing. Partition ratios are, however, only slightly different when measured at the usual and up to twice the usual sampling times. The height of the phase column in which partitioning is carried out is a factor in the efficiency of cell separation: higher columns give better separations but slower settling times. In contrast to the partitioning of macromolecules and small particulates the partitioning of cells is not particularly sensitive to cell size. Surface properties appear to be more important than surface area.

The separation of the components of a mixture by distribution between two immiscible liquids, either by bulk extraction or by liquid-liquid partition chromatography, is a well established technique. The aqueous two-phase systems that arise spontaneously when solutions of certain structurally dissimilar water-soluble polymers are mixed above critical concentrations are particularly useful for cell separations as they have high water contents, low interfacial tensions, and can be buffered and rendered isotonic (1-4). Cells distribute between one of the bulk phases and the horizontal interface on the basis of their surface properties.

Appropriate choice of concentration and composition of polymers and salts allows cell partitioning to be determined predominantly by charge-associated properties, non charge-related properties or receptor status. Cell separations often require multiple partitioning

---

[4]Current address: Department of Chemical Pathology, Royal Postgraduate Medical School, Hammersmith Hospital, DuCane Road, London W12 0NN, England

0097–6156/91/0464–0175$06.00/0
© 1991 American Chemical Society

steps as in, for example, a countercurrent distribution (CCD) procedure. A thin layer CCD (TLCCD) apparatus is generally used, designed to shorten the settling times of the biphasic systems to 5 to 8 min so that 60 transfer CCDs can be performed in about 5 to 8 hrs (5). CCD is used to separate cell populations and can also often subfractionate cell populations that are apparently homogeneous when examined by other physical separatory methods. Theoretical treatments of cell partitioning have considered it to be an equilibrium process. However studies of its kinetics reveal a dynamic process leading to the conclusion that cell separations and subfractionations by CCD actually depend on non-equilibrium conditions (6).

Cell Partitioning as a Dynamic Process

When an aqueous two-polymer phase system is mixed, an extensive interface is immediately formed; this develops into a series of PEG-rich and dextran-rich domains by coalescence. These domains are initially immobile but begin to move on reaching a critical size, usually within a few seconds; the PEG-rich areas generally move upwards and the dextran-rich areas downwards. Continuation of coalescence produces larger droplets and streams which proceed faster to the developing horizontal interface than the populations of smaller droplets, which have not coalesced to any degree. On coalescing with the horizontal interface these streams and large droplets contribute their contents to the developing top and bottom bulk phases, leaving the smaller droplets to persist in the bulk phases, moving relatively slowly to the horizontal interface with which they ultimately coalesce completing phase separation.

When cells are present in phase systems they influence phase separation by interaction with the interfaces; their partitioning behavior arises as a direct consequence. Microscopic examination of the phase systems as cell partitioning proceeds and also of samples removed at intervals shows that cells attach on the outer surface of dextran-rich droplets and streams that are ultimately delivered to the bottom phase and on the inner surface of PEG-rich droplets and streams that finally are delivered to the top phase. The degree of this interaction is dependent on cell type and phase composition (4,7-10). For example, in non charge-sensitive phases [i.e., systems without an electrostatic potential difference between the phases, (1,2)] of dextran (M.W. 500,000) - PEG (M.W. 6000) close to the critical point, erythrocytes of different species show cell-interface interactions in the order:

$$rat < turkey < human < pig < ox$$

Weakly associating cells (e.g., rat) are characterized by attaching to droplets with little change in cell shape, e.g., sitting proud on the outer surface of dextran-rich droplets (giving them a bumpy appearence); and sitting proud on the inner surface of PEG-rich droplets. They tend to readily associate and disassociate from droplets. There is little influence on the pattern of phase separation observed microscopically or quantitatively as measured by the rate of dextran depletion from the top phase. By contrast a strongly binding cell (e.g., pig) shows marked changes in cell shape on adhering to phase droplets, becoming extremely flattened and following the curvature of the droplet surface intimately. There is a marked effect on the pattern of phase separation, accelerating interactions between the dextran-rich domains and between PEG-rich domains with consequent increase in the rate of phase separation. Free cells are seldom seen. Increasing the polymer concentration increases the degree of association for all the cells so that changes from weak --> medium ---> strong are observed. Conversely, increasing the electrostatic potential difference ($\Delta\Psi$) between the phases by increasing the proportion of phosphate in isotonic phosphate-buffered saline or including PEG-ligands (e.g., PEG-palmitate) weakens these interactions. Cell partitioning into the top phase increases as these interactions are weakened. Such interactions have been quantitated by determining the wetting characteristics of the cell surfaces by the two phases by contact angle measurements (11, 20).

Because of the association of cells with phase droplets the distribution of cells in the phase system is intimately tied to phase separation. Figure 1 shows the effect of weakening cell-droplet interactions (at the indicated polymer concentrations) by increasing the $\Delta\Psi$ of the phase systems. In isotonic low phosphate/high NaCl system (low $\Delta\Psi$) human erythrocytes partition to the bulk interface and their partitioning into the top phase quickly reaches a low value. Increasing the indicated salt ratio increases $\Delta\Psi$ and weakens cell-droplet interactions thereby reducing the proportion of cells that are rapidly delivered to the horizontal interface.

A simple view of these time courses substantiated by kinetic analysis (Raymond, Gascoine and Fisher, unpublished) is that the curves consist of two components: a) a fast clearance curve, reflecting cells on rapidly moving streams, and b) a slow clearance curve, due to cells on smaller, slow settling droplets. It is this latter population which (together with any free cells) represents the majority of cells present in the top phase when the horizontal interface is clearly formed, and is measured as the cell partition ratio (quantity of cells in the top phase as a percentage of total cells added).

Implications of Cell Partitioning Dynamics for Cell Separations

Four general situations can be envisaged for the partitioning of a mixture of two cell populations, A and B, of which A attaches to phase droplets less strongly than B (Figure 2). Case 2 offers the possibility of cell separation by a few partition steps because cell population A partitions into the top phase while population B is at the interface. For separation by CCD the phase composition chosen is such that when the bulk interface is first clearly apparent, and therefore phase separation is almost complete, 20 to 80% of cells are still present in the top phase (case 3). As can be seen, case 1 will provide no separation as both cell populations are at the bulk interface and show no difference in partitioning. Similarly, there is no discrimination between cell populations in case 4 as nearly all the cells are in the top phase. It should be noted that, while the

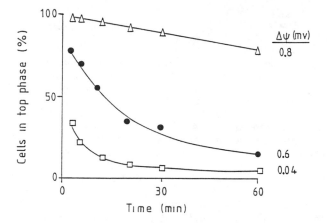

Figure 1.   Quantity of human erythrocytes partitioning into the top phase as a function of time in 5% dextran T500 (Pharmacia) and 4.5% PEG 6000 (BDH) biphasic systems with different electrostatic potential differences between the phases ($\Delta\psi$): 0.01 M sodium phosphate buffer, pH 6.8, + 0.15 M NaCl (0.04 mv); 0.068 M sodium phosphate buffer, pH 6.8, + 0.075 M NaCl (0.6 mv); 0.09 M sodium phosphate buffer, pH 6.8, + 0.03 M NaCl (0.8 mv). A qualitative measure of $\Delta\psi$ was obtained using 3M KCl-agar salt bridges and an electrometer.

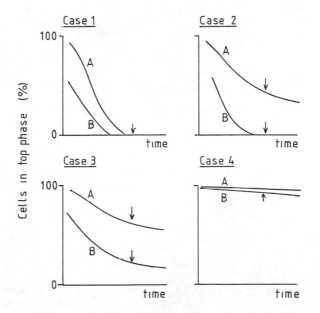

Figure 2. Schematic representations of possible time courses of cell partitioning into the top phase. A and B are two cell types which differ in their attachment to bottom phase droplets, cells A showing weaker attachment. The arrows indicate the time at which the horizontal interface between the phases becomes sharp, the operational condition at which sampling or transfer steps in CCD are usually performed.

partition ratio is a result of the dynamics of the early events in cell partitioning, its measured value will generally not be greatly affected by extending (e.g., doubling) the time beyond that required for phase separation to be virtually complete (Figure 3). Because the partition ratio is determined by the dynamics of phase separation which themselves are influenced by a number of parameters (e.g., tie-line length, temperature, vessel geometry), manipulation of the latter may yield improved cell separations (12).

Effect of Speed of Phase Separation on Cell Partitioning

The Albertsson TLCCD apparatus (5) uses phase columns approximately 3.5 mm high. Such thin layers of phases separate more rapidly than do phases in the classical Craig CCD unit in which the tubes are arranged vertically and have heights of about 15 cm. Rapidity of separation is often desirable, particularly when labile biomaterials are being processed. However, with cells, in which partitioning depends on cell-phase droplet interactions, the speed of phase settling can affect cell separation. Walter and Krob (12) recently found that the partition ratio of cells in low phase i.e., short columns (using horizontally arranged tubes) is higher than when the cells settle in high phase i.e., long columns (vertically arranged tubes). Furthermore, in a model system of rat erythrocytes in which subpopulations of red blood cells of different but distinct ages were isotopically labeled (12), low phase columns were found to give less efficient cell fractionations than high phase columns. Because in a given phase system the partition ratio for cells was always lower in the vertical than in the horizontal settling mode, the greater efficiency of separation in high phase columns could reflect the smaller number of cells which partition into the top phase. By using phase compositions which produced a similar percentage of added cells in the top phase in vertical and in horizontal settling modes it has been shown that the higher efficiency in the vertical mode is not dependent on the quantity of cells. The most likely reason for these results is that increasing the speed of phase settling (as with low phase columns) may remove the droplets of one phase suspended in the other more rapidly than cells can attach to them, thereby interfering with the mechanism whereby cells partition.

Effect of Short Settling Times on Cell Separations

In cases 1, 2 and 3 (Figure 2) separations of A and B can also be obtained if CCD transfers are made at times earlier than those required for nearly complete phase separation. Little attention has thus far been paid to using such short settling times. Eggleton, Fisher and Sutherland have built a hand-held, hand-operated modified 30 chamber version (13) of the Pritchard device (14,15), which provides "thick" layers of phase (i.e., high phase columns) in contrast to the conventional thin-layer CCD (TLCCD) apparatus (1,5,16). Short settling times, 105 sec, gave good separations at least in a model system consisting of cells having significantly different percentages of added cells in the top phase (e.g., rat and sheep erythrocytes) (Figure 4). Short settling times have also been used to advantage in a study of the surface charge-associated heterogeneity.

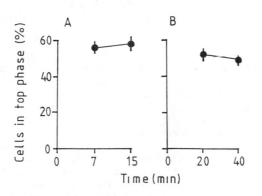

Figure 3. The effect of doubling the "usual" sampling time on the percentage of added human erythrocytes in the top phase. A. A 5% dextran T500 (Pharmacia) and 3.5% PEG 8000 (Union Carbide) in 0.15 M NaCl + 0.01 M sodium phosphate, pH 6.8, was permitted to settle horizontally for 7 or 15 min before being brought to the vertical position for 1 min prior to sampling. B. A 5% dextran T500 and 4% PEG 8000 in 0.11 M sodium phosphate, pH 6.8, was permitted to settle vertically for 20 or 40 min.

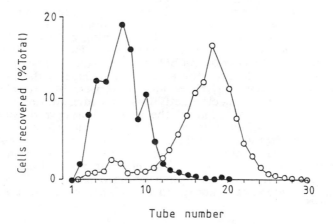

Figure 4. Separation of a mixture of sheep (●) and rat (○) erythrocytes in a hand-held, hand-operated CCD unit, a modified 30 chamber version (13) of the Pritchard device (14,15) with "thick" layers of phase. A 5% dextran T500 (Pharmacia) and 4.5% PEG 6000 (BDH) non charge-sensitive phase system (containing 0.15 M NaCl + 0.01 M sodium phosphate, pH 6.8) was used with a 15 sec mixing and a 105 sec settling time.

The effect of prolonged settling times (up to two hr), in high and low phase columns, on the measured cell partition ratios and on the separability of cell populations is currently being explored (18).

Closely Related Cell Populations.   By using a model system consisting of rat erythrocytes in which subpopulations of red blood cells of distinct ages were labeled isotopically it was found that partitioning proceeds over the entire time period examined (2 hr) as evidenced by the continuous change in relative specific activity of cells in the top phase (a measure of the separation of labeled and unlabeled cells) as the partition ratio falls (e.g., Figure 5).   In control sedimentation experiments in top phase there is almost no change in the quantity of cells present when vertical settling is used and no separation of specific subpopulations is found.

In the horizontal settling mode the initially higher cell partition ratio, as compared to vertical settling, decreases to a greater extent with longer times.   And, although the efficiency of separating a cell subpopulation also increases with time, a given purity of cells is only achieved at a lower partition ratio than in the vertical settling mode. Cell sedimentation in top phase is appreciable over 2 hrs in the horizontal settling mode but does not result in the separation of cell subpopulations.

Cell Populations with Different Partition Ratios and Sizes.   With model systems of (a) ^{51}Cr-labeled rat red cells mixed with an excess of human erythrocytes and (b) ^{51}Cr-labeled sheep erythrocytes mixed with an excess of human red blood cells the effect of relative cell partition ratios and sizes in high and low phase columns on the efficiency of separation was examined. Rat and sheep red cells are both appreciably smaller than the corresponding human cells. Rat red cells have higher while sheep red cells have lower partition ratios than do human erythrocytes. As before, the partitioning process was again found to proceed over the entire 2 hr period of the experiment and, with vertical settling, there was no appreciable contribution to the change in relative specific activities by sedimentation. However, the more rapid sedimentation of the larger human red cells had a measurable effect on the relative specific activities obtained during cell partitioning in the horizontal mode: enhancing the rat-human cell separation and diminishing the sheep-human cell separation.

Partitioning cells in high phase columns would thus appear to be of advantage with respect to increasing separation efficiency and decreasing the influence of non-partition related parameters (e.g., cell size effects).

Effects of Cell Surface Area on Cell Separations

Albertsson's generalized application of the Brønsted equation (1) states that the partition coefficient, $\underline{K}$, which is C_T/C_B, depends on e $^{\lambda M/kT}$, where C_T and C_B are the partitioning material's concentration in the top and bottom phases, respectively, M is the molecular weight, k the Boltzmann constant and T the absolute temperature. λ is a factor which depends on properties other than

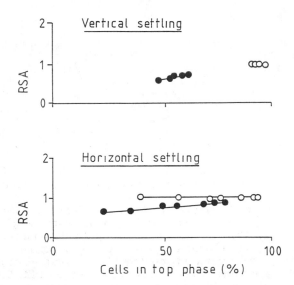

Figure 5. The effect of the mode (i.e., high or low phase columns) of partitioning (●) and sedimentation (○) on the efficiency of cell fractionation in a dextran-PEG biphasic system as exemplified by rat erythrocytes containing an isotopically (^{59}Fe) labeled subpopulation of 43-day old cells. The lower the relative specific activity, RSA [defined in (12)], when fractionating erythrocyte populations containing such old cells, the greater is the efficiency of fractionation. Each point represents a time sample for vertical settling (high phase column), 20 min to 2 hrs, and for horizontal settling (low phase column), 7.5 min to 2 hrs, increasing from right to left. The continuous change in the RSA value over the 2 hr period of the experiment indicates that partitioning continues over this entire time period. See text for additional discussion.

molecular weight (i.e., interactions between the surface of the material and the two phases). Albertsson has further proposed that for particulates M should be replaced by A, the particle's surface area (1). Thus size and surface properties are considered to determine the partitioning behavior of biomaterials in a most sensitive manner since these parameters are exponentially related to K. Boucher (19) has questioned the appropriateness of applying the Brønsted equation to partitioning phenomena and has formulated a Boltzmann-type expression in which an exponential relationship exists between parameters including surface area contributing to the partitioning behavior and K value.

The partitioning of biomaterials between bulk phases (as in the case of macromolecules) switches to partitioning between one phase and the interface (as in the case of particles) in a size dependent manner (20). The partitioning of a particle to the interface would be expected to occur on theoretical considerations at a particle diameter of about 0.03 μm in phase systems having interfacial tensions of the order associated with a composition of 5% (w/w) dextran T500/4% (w/w) PEG 8000 (20). While a relationship between molecular mass and K has been found in the case of macromolecules and very small particulates (e.g., small viruses) (see ref.1) the dependence of partitioning behavior on surface area in the case of larger particulates like cells has never been experimentally established.

Unilamellar phospholipid vesicles of identical composition but different sizes provide good model systems with which to explore this question (21,22). In studies with negatively charged vesicles of phosphatidylglycerol-phosphatidylcholine (4:6 molar ratio) and phosphatidic acid-phosphatidylcholine (4:6 molar ratio) in charge sensitive phase systems (5% dextran T500 and 5% PEG 6000 in 0.11M sodium phosphate, pH 6.8) there is a marked decrease in log K with increasing vesicle surface area (up to about $0.1 \mu m^2$), whilst larger vesicles showed the same log K irrespective of size [21]. Biological cells differ both in size and composition so that it is not possible to examine the effect of size independently of composition. An examination of some partitioning data that are in the literature for what they reveal about the involvement of size (or surface area) in contributing to the partitioning of cells is informative. Log K of erythrocytes from different species (volumes ranging from 35-135 μm^3) in charge-sensitive phases correlate with the cells' relative electrophoretic mobilities (23). The latter depend on the charge per unit area and not on cell size. Log K values for erythrocytes from different species partitioned in non charge-sensitive phases (24) or by hydrophobic affinity partition with PEG-palmitate (25) correlate with the cells' membrane ratio of poly/monounsaturated fatty acids, and to the membranes' relative composition of phosphatidylcholine and sphingomyelin, parameters which also do not pertain to total surface area.

Thus size can be one of the determinants of the K value of biomaterials in the smaller size ranges (e.g., macromolecules, viruses and small vesicles), but surface concentration of particular constituents (lipids, charged glycoproteins) outweigh total surface area per se in the determination of the partition of large particulates (e.g., cells).

Effects of Phase Composition on Cell Separation

Properties of phase systems such as interfacial tension, electrostatic potential difference between the phases ($\Delta\psi$), viscosity and rate of phase separation are critically dependent on the composition of the phase system (26) and the temperature [e.g., increase in polymer concentrations produce marked increases in interfacial tension and in the viscosity of the dextran-rich phase (in dextran-PEG systems)]. The sensitivity of these properties to the molecular weight of the polymers has been well described (1, 2, 4, 26). As these properties affect cell partitioning, often with an exponential dependency, there is ample opportunity for the sensitive manipulation of partitioning to optimize cell separations. However, such sensitivity has an operational drawback in that variation in performance of phase systems is experienced with different batches of polymers. Success depends upon careful preparation and use of phase systems (34). Defining phase compositions and citing polymer batch numbers gives some characterization of the phase used. Measurements of interfacial tension, $\Delta\psi$, density difference, and viscosity provide improved characterization but are not easily made. The partitioning of bioparticles suffers from the lack of a quantative theoretical basis. A thermodynamic theory has been developed by Sharp, see (20), for the partition of particles which predicts an exponential relationship between the partition coefficient and the degree to which one of the phases preferentially wets the particle surface (described by the contact angle between the particle and the phase boundary), and the interfacial tension. Although the general features of this relationship have been confirmed by experiment, the numerical coefficients are orders of magnitude different from the predicted values. The dynamic nature of cell partitioning described above is the source of this discrepancy. The adherence of particles to phase droplet surfaces, which is thermodynamically determined by wetting characteristics, is opposed by the particle being swept off the surface by the turbulent flow of the droplet, as phase separation proceeds. This latter is an uncontrolled process. It results in partition coefficients being dependent on the conditions and rate of phase separation, as already discussed. Although advantage can be taken of this in the design of settling chambers, such an approach is empirical.

Biospecific Cell Separations: Affinity Partitioning

A dramatic increase in the selectivity of cell partitioning can be achieved by combining it with biospecific affinity methods. Cells expressing antigen can be specifically partitioned into the top PEG-rich phase by treating the cells with the corresponding antibody coupled to PEG (27-33). "Double-layer" methods have been reported using secondary antibodies (29) and protein A (30,31) covalently linked to PEG. PEG-avidin used with biotinylated antibody has discriminated between subsets of cells whose number of antigen binding sites differs only by a factor of three (31). Distinctive reactivities of the surface of a desired cell subpopulation (e.g., paroxysmal nocturnal hemoglobinuria red cells) have been used to advantage to alter the cells' partition ratio and effect their extraction (34).

Despite these encouraging results, the challenge for cell affinity partitioning is to extract small subpopulations of cells with low antigen/receptor status. These are often the cells of greatest biological interest. Biospecific affinity cell partitioning has yet to achieve this. New methods for amplifying the affinity signal and possible new methodologies for CCD are under study.

Acknowledgments

DF and FDR thank the Royal Free Hospital School of Medicine and the Peter Samuel Royal Free Fund. Work in HW's laboratory was supported by the Medical Research Service of the Department of Veterans Affairs.

Literature Cited

1. Albertsson, P.-Å. Partition of Cell Particles and Macromolecules; Wiley-Interscience: New York, 1986.
2. Partitioning in Aqueous Two-Phase Systems. Theory, Methods, Uses, and Applications to Biotechnology; Walter, H.; Brooks D.E.; Fisher, D., Eds.; Academic Press: Orlando, 1985.
3. Separations Using Aqueous Phase Systems. Applications in Cell Biology and Biotechnology; Fisher, D.; Sutherland, I.A., Eds.; Plenum Press: New York, 1989.
4. Walter, H.; Johansson, G. Anal. Biochem. 1986, 155, 215-242.
5. Albertsson, P.-Å. Anal. Biochem. 1965, 11, 121-125.
6. Fisher, D.; Walter, H. Biochim. Biophys. Acta 1984, 801, 106-110.
7. Raymond F.D. The Partition of Cells in Two Polymer Aqueous Phase Systems; PhD Thesis; University of London: London, 1981.
8. Raymond, F.D.; Fisher, D. Biochim. Biophys. Acta 1980, 596, 445-450
9. Raymond, F.D.; Fisher, D. Biochem. Soc. Trans. 1980, 8, 118-119.
10. Raymond, F.D.; Fisher, D. In Cell Electrophoresis in Cancer and Other Clinical Research; Preece, A. W.; Light,P.A. Eds.; Elsevier/North Holland Biomedical Press: Amsterdam, 1981; pp 65-68.
11. Youens, B.N.; Cooper, W.D.; Raymond, F.D.; Gascoine, P.S.; Fisher, D. In Separations Using Aqueous Phase Systems. Applications in Cell Biology and Biotechnology; Fisher,D.; Sutherland, I.A., Eds.; Plenum Press, New York, 1989; pp 271-280.
12. Walter, H.; Krob, E.J. Biochim. Biophys. Acta 1988, 966, 65-72.
13. Eggleton, P.; Sutherland, I.A.; Fisher, D. In Separations Using Aqueous Phase Systems. Applications in Cell Biology and Biotechnology; Fisher,D.; Sutherland, I.A., Eds.; Plenum Press, New York, 1989; pp 423-424.

14. Pritchard, R.N.; Halpern, R.M.; Halpern, J.A.; Halpern, B.C.; Smith., R.A. Biochim. Biophys. Acta 1975, 404, 289-299.
15. Sutherland, I.A. In Partitioning in Aqueous Two-Phase Systems. Theory, Methods, Uses, and Applications to Biotechnology; Walter, H.; Brooks, D.E.; Fisher, D. Eds.; Academic Press: Orlando; 1985, pp 149-159.
16. Treffry, T.E.; Sharpe, P.T.; Walter, H.; Brooks, D.E. In Partitioning in Aqueous Two-Phase Systems. Theory, Methods, Uses, and Applications to Biotechnology; Walter,H.; Brooks,D.E.; Fisher, D. Eds.; Academic Press: Orlando; 1985, pp 131-148.
17. Crawford, N.; Eggleton, P.; Fisher, D. In this volume.
18. Walter, H.; Krob, E.J.; Wollenberger, L., submitted for publication.
19. Boucher, E.A. J. Chem. Soc., Faraday Trans. 1 1989, 85, 2963-2972.
20. Brooks, D.E.; Sharp, K.A.; Fisher, D. In Partitioning in Aqueous Two-Phase Systems. Theory, Methods, Uses, and Applications to Biotechnology; Walter, H.; Brooks, D.E.; Fisher, D. Eds.; Academic Press: Orlando; 1985, pp 11-84.
21. Walter, H.; Fisher, D.; Tilcock, C. FEBS Lett., 1990, 270, 1-3.
22. Tilcock, C.; Cullis, P.; Dempsey, T.; Youens, B.N.; Fisher,D. Biochim. Biophys. Acta 1989, 979, 208-214.
23. Walter, H.; Selby, F.W.; Garza, R. Biochim. Biophys. Acta 1967,136, 148-150.
24. Walter, H.; Krob, E.J.; Brooks, D.E. Biochemistry 1986, 15, 2959-2964.
25. Eriksson, E.; Albertsson, P.-Å.; Johansson, G. Mol. Cell. Biochem. 1976, 10, 123-128.
26. Bamberber,S.; Brooks,D.E.; Sharp,K.A.; Van Alstine, J. and Webber, T.J. In Partitioning in Aqueous Two-Phase Systems. Theory, Methods, Uses, and Applications to Biotechnology; Walter, H.; Brooks, D.E.; Fisher, D. Eds.; Academic Press: Orlando; 1985, pp 85-130
27. Karr, L.J.; Shafer, S.G.; Harris, J.M.; Van Alstine, J.M.; Snyder, R.S. J. Chromatogr. 1986, 354, 269-282.
28. Sharp, K.A.; Yalpani, M.; Howard, S.J.; Brooks, D.E. Anal. Biochem. 1986, 154, 110-117.
29. Stocks, S.J.; Brooks, D.E. Anal. Biochem. 1988, 173, 86-92.
30. Karr, L.J.; Van Alstine, J.M.; Snyder, R.S.; Shafer, S.G.; Harris, J.M. In Separations Using Aqueous Phase Systems. Applications in Cell Biology and Biotechnology; Fisher,D.; Sutherland, I.A., Eds; Plenum Press, New York, 1989;pp 193-202.
31. Karr, L.J.; Van Alstine, J.M.; Snyder, R.S.; Shafer, S.G.; Harris, J.M. J. Chromatogr. 1989 442, 219-217.
32. Stocks, S.J.; Brooks, D.E. (1989).In Separations Using Aqueous Phase Systems. Applications in Cell Biology and Biotechnology; Fisher, D.; Sutherland, I.A., Eds.; Plenum Press, New York, 1989; pp 183-192.

33. Delgado, C.; Francis, G.E.; Fisher, D. In Separations Using Aqueous Phase Systems. Applications in Cell Biology and Biotechnology; Fisher, D.; Sutherland, I.A., Eds.; Plenum Press, New York, 1989; pp 211-213.
34. Pangburn, M. K.; Walter, H. Biochim. Biophys. Acta 1987, 902, 278-286.

RECEIVED April 15, 1991

ELECTROPHORETIC AND MAGNETIC METHODS

Chapter 13

Population Heterogeneity in Blood Neutrophils Fractionated

by Continuous Flow Electrophoresis (CFE) and by Partitioning in Aqueous Polymer Two-Phase Systems (PAPS)

Neville Crawford[1], Paul Eggleton[1], and Derek Fisher[1,2]

[1]Department of Biochemistry and [2]Molecular Cell Pathology Laboratory, Royal Free Hospital School of Medicine, University of London, Hampstead, London NW3 2PF, England

Isolated human blood neutrophils have been fractionated into subpopulations by CFE and PAPS (Dextran and PEG). Although CFE seems to be the more innocuous procedure with respect to cell viability and functions, both techniques give similar separation profiles, i.e. roughly gaussian and extending over 15–20 fractions. The electrokinetic basis for the separation has been established by analytical cytopherometry of the fractions. Functional investigations which have included chemotaxis, phagocytosis of bacteria, adhesion to surfaces and stimulated respiratory burst showed in both CFE and PAPS–separated cells an inverse relationship between surface membrane electronegativity and functional and metabolic competence. This relationship is being explored by analysis of membrane constituents, receptor status for activating ligands, priming effects, post–receptor signalling events and the dynamics of the membrane/cytoskeletal axis during motile activities. Neutrophil heterogeneity may be important in margination in selective recruitment to inflammatory foci, hereditary motility defects and myeloproliferative diseases, as also in neutrophil interactions with natural and foreign surfaces.

Circulating neutrophils are relatively quiescent cells, but during an inflammatory response they show remarkable changes in morphology, metabolism and functional expressions. Many of these functional events such as cell–cell adhesion, chemotaxis, phagocytosis and oxidative burst, are associated with rapid changes in surface membrane topography which include polarization, expression of receptors (or their upregulation), changes in shape, formation of pseudopodia, modifications to surface glycoproteins and both intra and extracellular secretion of components stored in granules. For these functional expressions the cell surface membrane may undergo both widespread and domain–restricted alterations in biophysical, biochemical and electrokinetic properties. The role neutrophils play in the body's defense mechanisms may relate to their capacity to make these changes.

0097–6156/91/0464–0190$06.00/0

The total life span of a neutrophil in the body is almost 2 weeks, but a greater part of this time is spent maturing in the bone marrow. Estimates of their mean half life in the circulating pool have varied between just a few hours and one day, depending upon the type of label used. These aspects of leucopoeiesis are, however, complicated by the presence of an intravascular pool of neutrophils which are reversibly adherent to vessel walls and termed "marginated neutrophils". This marginated pool of cells which is in equilibrium with the circulating neutrophil pool, can account for as much as 70% of the total intravascular content of neutrophils. Certain endogenous agents (e.g. adrenaline, prostacyclin etc.) can disturb this equilibrium by decreasing neutrophil adhesion to vessel wall cells and causing transient leukocytosis. In acute inflammatory states, other mediators are released, which can affect the integrity of the endothelial cell layer and allow marginated neutrophils to migrate out of the blood vessel and to "home in" on tissue sites of inflammation. The granule and lysosome-stored enzymes that these stimulated migratory neutrophils release in transit can be damaging to tissues, but other release products can actually take part in tissue repair processes. Like monocytes and macrophages, neutrophils are also phagocytic and, during particle ingestion, a similar range of vesicle-stored enzymes is discharged intracellularly into the phagocytic vesicles to facilitate breakdown of internalized foreign materials.

All these migratory, secretory and phagocytic functions of neutrophils are triggered by the cell's associations with particles or surfaces, and, whilst some of these interactions are determined by characterised receptor-ligand binding events, others depend on a less specific expression of adhesion forces determined by the hydrophobic and electrostatic properties of surface oriented plasma-membrane constituents.

During the last 10 years or so, there have been a number of reports suggesting that circulating neutrophils are heterogenous in both membrane analytical properties and in functional competence. In a clinical context the importance of such heterogeneity lies in the prospect of identifying subpopulations of neutrophils in the circulating pool more or less attenuated to respond to inflammatory stimuli. This has led to an interest in neutrophil priming, i.e. the effect of cytokines and other agents which in themselves display no stimulatory effects but amplify the subsequent action of conventional stimuli. Other neutrophils may also be present which may be functionally impotent, either because they are immature, haven't yet been primed, or they are simply effete and destined for early removal by the reticuloendothelial system (RES). An understanding of the basis of such neutrophil heterogeneity would not only be of value comparatively in the study of leucocytes in various diseases, but new targets may be revealed which would allow a more rational approach to drug design in the search for agents which interfere with or amplify certain neutrophil functions which might be advantageous in the treatment of disease.

In 1975 Gallin and colleagues observed that neutrophil activation by chemotactic factors was accompanied by a reduction in the cell's surface electrical charge (1). Later this group also showed a similar inverse relationship between neutrophil surface electronegativity and their response to degranulating stimuli (2). Earlier Weiss et al. (3) had shown that the removal of surface sialic acid groups from monocytes and macrophages by neuraminidase resulted in enhanced phagocytic activity and in a similar study Lichtman & Weed (4) reported that mature granulocytes from the bone marrow with membranes rich in sialic acid were more rigid, less adhesive and showed poorer phagocytic capacity than less electronegative circulating blood granulocytes. In a study of the role of surface charge in phagocytosis Nagura and his colleagues (5) showed that the lowering of the surface electronegativity of macrophages by treatment with protamine sulphate substantially enhanced their phagocytic activity.

Using a surface charge dependent separation of porcine neutrophils by continuous flow electrophoresis and applying to the pooled fractions a quantitative radioactive particle assay for phagocytic behaviour, Spangenberg & Crawford (6) reported that the subpopulations of the lowest surface electronegativity showed the greatest ability to phagocytose. These findings, though supporting the earlier work of Gallin (1,2) and others (3 – 5) differed from those of Walter and co–workers (7) using partitioning in a charge–sensitive two–phase aqueous polymer (Dextran/polyethylene glycol) system to fractionate another species of professional phagocytes, the monocytes. They observed that an increase in the ability to internalise particles by phagocytosis correlated with an increase in cell fraction partitioning coefficients although actual electrophoretic mobility values were not measured in these experiments.

In order to further investigate the relationship between membrane electrokinetic properties and function in phagocytic cells and to explore this apparent conflict between two charge sensitive separation procedures we have been studying human neutrophils isolated from the circulating pool. They have been separated into heterogeneity profiles by both continuous flow electrophoresis [CFE] and by partitioning in a charge sensitive aqueous polymer two–phase system with counter current distribution [PAPS/CCD]. We reasoned that although surface membrane electrical charge was almost certainly a major operational parameter in both separation techniques, the surface of a neutrophil, unlike red cells and synthetic particles, is highly ruffled, and the presence of invaginated domains may result in different electrokinetic characteristics in the two systems. Separations by CFE are based almost entirely upon the disposition and contribution of charged membrane moieties to the overall "zeta potential" of the cell surface, whereas in the charge sensitive PAPS procedures the efficacy of particle partitioning is influenced by both the interfacial tension and the electrostatic potential difference between the polymer phases.

Materials, Apparatus and Methods

Neutrophil Isolation from Blood. All fresh blood samples were taken from healthy donors into EDTA anticoagulant (1.5 mg/ml). Later in the studies fresh blood from the Transfusion Service Laboratories anticoagulated with ACD adenine was used. To our knowledge no donor had any condition that included an inflammatory response. There was no detectable difference between neutrophils isolated with the different anticoagulants. At the outset of these studies it was decided that for a proper comparative evaluation of the two separation systems a rapid, high yield, high recovery, non–damaging, neutrophil isolation procedure was required that preferably would also avoid any prior contact with polymer materials likely to become associated with the cell surfaces and influence fractionation or functional properties. After numerous trials with a range of conventional procedures using a variety of commercially available polymer materials a differential centrifugation procedure (8) was selected as fulfilling most of the required criteria for these studies. Table I shows the yield and recovery details for this differential centrifugation procedure compared with the values for other more conventional procedures, i.e. sedimentation in Dextran and Dextran/Ficol–Hypaque.

Table I. Comparative Neutrophil Yield and Purity After Isolation[a]

Extraction Procedure	Neutrophil Yield (%)	Neutrophil Purity (%)
Differential centrifugation	70–80	90
Ficol–Hypaque	50	90–95
Dextran sedimentation	50	80

[a]Based on at least three separate experiments.

Figure 1 shows the heterogeneity profiles from TAPP/CCD separations of the same human neutrophils isolated by differential centrifugation, Dextran sedimentation and a Ficol/Hypaque procedure.

The CFE and PAPS CCD Apparatus. For continuous flow electrophoresis the ELPHOR VAP 22 instrument of Bender & Hobein (Munich) was used. This model has a separation chamber of 0.5 mm buffer film depth between the plates and 90 exit ports across a 10 cm. wide chamber. The electrode buffer used was triethanolamine/acetic acid [100 mM triethanolamine adjusted to pH 7.4 with glacial acetic acid] conductivity ~3500 μs/cm and 180 m osm/Kg and the separation chamber buffer was a 10 fold dilution of this buffer to which was added 280 mM glycine and 30 mM glucose (conductivity ~650 μs/cm and 320 m osm/Kg).

Routinely the chamber buffer flow rate of the VAP 22 was set at ~2.5 ml/hr/exit port and the flow rate for the sample input pump set at 2.0 ml/hr. The optimum electrical parameters for electrophoresic separation of human neutrophils were 150 mamps at 110 V/cm, and the chamber was maintained at a constant temperature of 8°C during the fractionation. For each electrophoretic run the chamber was first washed with a dilute Teepol solution and rinsed with many chamber volumes of deionised water. The chamber walls were then pre–coated by filling the chamber with a 3% albumin solution, and after 1 hour this was washed away with many volumes of distilled water before filling the chamber with separation buffer. This albumin coating reduces the zeta potential of the glass chamber walls and improves the cell separation. Electrophoretic distribution profiles were constructed after counting the cell fractions in a Coulter counter [Model ZB1]. For routine studies turbidometric measurements at 500 nm were used to define the profile. For both the CFE and CCD separations, analytical and functional studies were made on the individual fractions wherever possible. If pooling was necessary the profile was divided into three or four pools of equal cell numbers as determined by Coulter counting.

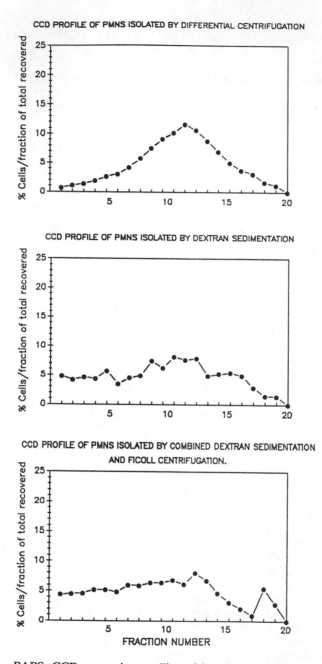

Figure 1. PAPS–CCD separation profiles of human neutrophils isolated from the same blood sample by three different procedures. Profile distortion can occur if the cells are exposed to polymer materials before CCD fractionation. In this two–phase system cell surface electronegativity is considered to increase from left to right.

The PAPS counter current distribution (CCD) apparatus used was a hand manipulated instrument (designed and developed by Drs I. Sutherland (M.R.C., Mill Hill, London) and D. Fisher (Royal Free Hospital School of Medicine, London) consisting of 30 cavities between 2 circular plexiglass plates. The bottom cavity was of 0.71 ml volume but since with the Dextran/PEG system chosen the cells partitioned between the top phase and the interface, only 0.65 ml volume of Dextran was used to allow the interface to locate with the Dextran rich bottom phase. In all experiments described here a charge–sensitive phase system of 5% Dextran, 6.5% polyethylene glycol in 0.11M sodium phosphate buffer (pH7.4) was used. Generally after 15 inversions to mix the phases, a settling time of 2.5 mins before moving to the next cavity position and 20 successive transfers, the cells were ready for collection. Distribution profiles were constructed after counting the number of cells in each of the 20 fractions in a Coulter counter, as with the CFE separations.

Analytical electrophoretic mobility measurements were made with a cylindrical or rectangular chamber Zeiss cytopherometer and EPM values were expressed as μm/sec/V/cm.

Procedures for Functional Studies. The technique for the measurement of chemotaxis was based upon that developed by Nelson et al. (9). This method measures the migration of neutrophils under agarose and the chemotactic agents used were endotoxin, FMLP and AB serum. For a quantitative expression of phagocytic activity, opsonised bacteria (*Staphylococcus aureus*) were used at a bacteria/neutrophil ratio of 5:1. After 15 min phagocytosis at 37°C, extracellular and surface adherent bacteria were lysed with lysostaphin (20 units/ml), and slides of the cells were prepared by cytospin and stained with Prodiff-1 and 2. The % of cells showing ingested bacteria (Phagocytic %) and the mean number of bacteria/phagocytosing cell (Phagocytic index) were determined on at least 100 cells.

Superoxide anion production in the neutrophil subpopulations was determined by a semi–quantitative microscopic nitroblue tetrazolium [NBT] procedure or by a quantitative NBT assay using microtitre plates (10). Both basal (spontaneous) and FMLP stimulated NBT determinations were made on all the cell fractions.

Results and Discussion

In order to establish the electrokinetic characteristics of a PAPS/CCD neutrophil fractionation profile, analytical cytopherometry measurements of electrophoretic mobilities (EPM) were made on four pooled fractions of equal cell number (quartiles) across the separation profile. Figure 2 shows two typical CCD separations of cell fractions (1–20, with increasing partition coefficient) with the EPM values for the fraction pools P1 to P4 and that of the total neutrophil population before PAPS/CCD separation. For both subjects these CCD separations correlated with cell surface electronegativity as reflected in the electrophoretic mobility measurements. Fractions having the lowest electrophoretic mobility contained cells with the lowest partition coefficients and *vice versa*.

In a study with neutrophils from 3 different subjects separated by CFE, analytical cytopherometry of pooled fractions (quartiles) revealed a similar correlation. Figure 3 shows the analytical EPM results as mean ± standard

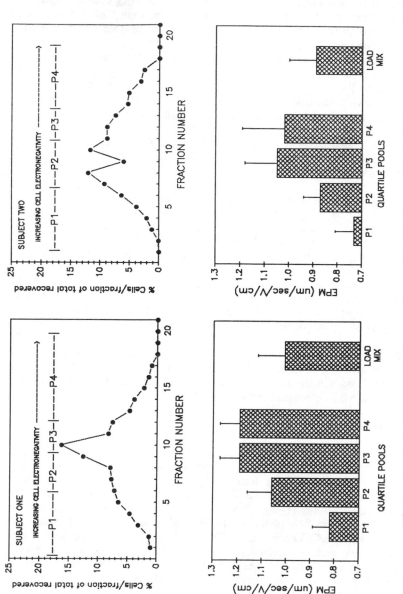

Figure 2. Typical PAPS–CCD separation profiles for neutrophils from two normal subjects. The histograms below show analytical electrophoretic mobility values for the quartile pools P1, P2, P3 and P4 and for the total cell population before CCD fractionation.

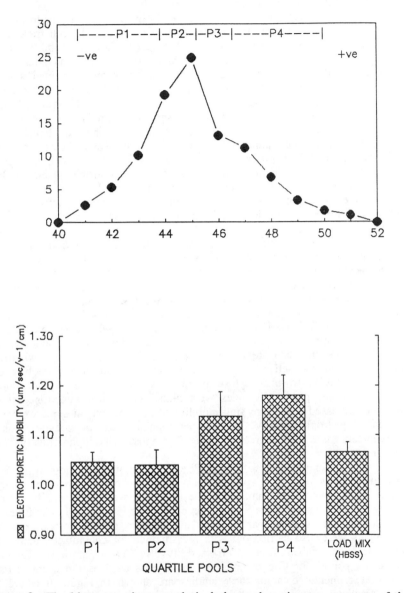

Figure 3. The histogram shows analytical electrophoresis measurements of the mobilities of neutrophil pools (quartiles) taken from CFE separations of neutrophils from 3 normal subjects (Means ± SD). Above is a typical CFE separation profile of the cells from one of the subjects.

deviations for the neutrophils from 3 different normal subjects together with a typical CFE profile of the cells from one of the subjects.

To seek a molecular basis for these mobility differences, neutrophils from two normal subjects were electrophoretically separated, and the cell fractions were again pooled into four quartile fractions. The electrophoretic mobilities of these fractions were measured by analytical cytopherometry before and after treatment with neuraminidase to remove neuraminidase–labile surface oriented sialic acid moieties (0.4 units Sigma Co. affinity purified neuraminidase in 1 ml for 1 hour at 37°C). These results are presented in Figure 4 where it can be seen that in both studies control pooled fractions across the separation profile increased in EPM towards the anodal side of the chamber, but they still showed significant EPM differences after the neuraminidase treatment, although EPM decreased in all cases. Evidently the carboxyl groups of sialic acid present on integral membrane sialoglycoproteins are major contributors to the neutrophil's electrophoretic mobility, but other ionogenic groups may also be influential.

In order to explore the relationship between neutrophil surface membrane electronegativity and function, and membrane–mediated motile activities in particular, both phagocytic and chemotactic activities were studied across the continuous flow electrophoresis and the PAPS/CCD separation profiles and phagocytic index were measured. Two approaches were used in CFE separations, (1) phagocytic parameters were measured on the pooled quartiles after CFE separation by quantitative microscopy of cytospin preparations and (2) cells were allowed to phagocytose bacteria before they were separated by CFE and then pooled into fraction quartiles for collection by cytospin and microscopic examination. The four pools of equal numbers of cells were designated P1 to P4 with P1 representing the least and P4 the most electrophoretically mobile cells.

Figure 5 shows the data from these experiments. Irrespective of whether the neutrophil phagocytic activity was measured after separation and pooling or the cells were allowed to phagocytose bacteria before CFE separation, both the phagocytic % and phagocytic index were higher in the least electronegative cell pools and decreased with increasing mobility of the fraction. In some experiments the least electrophoretically mobile neutrophils had phagocytic capacities approximately two fold greater than the most electrophoretically mobile neutrophils.

Similar findings emerged from PAPS/CCD separations of normal neutrophils followed by measurements of phagocytosis. Here the pooled fractions containing neutrophils of lower EPM values, e.g. the P1 pools, were substantially greater in phagocytic competence expressed either as phagocytic % or phagocytic index, than the fractions of greater partitioning and higher electrophoretic mobilities. Experiments were also carried out with measurement of chemotactic activities on pooled fractions taken from CFE and PAPS/CCD separations. Either the chemotactic peptide or C3b complement factor was used as the chemoattractant. Typical experimental data for these studies are shown in Figure 6 where, once again, fractions having the lowest electrophoretic mobilities have greater chemotactic indices than the more electronegative cells, confirming, as with phagocytosis, an inverse relationship between surface electronegativity and these particular membrane mediated motile functions. Other data (not recorded here) showed that in the measurement of respiratory burst, assayed as the amount of NBT reduced by resting and FMLP–stimulated neutrophils, both the low basal and higher stimulated values were always greater in the least electronegative fractions

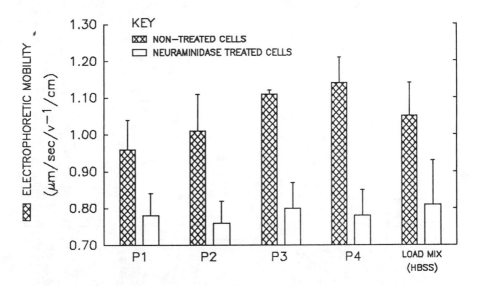

Fig 4. Electrophoretic mobility measurements of pooled neutrophil fractions (quartiles) from CFE separations of cells from two normal subjects. Measurements were made before and after treatment with neuraminidase.

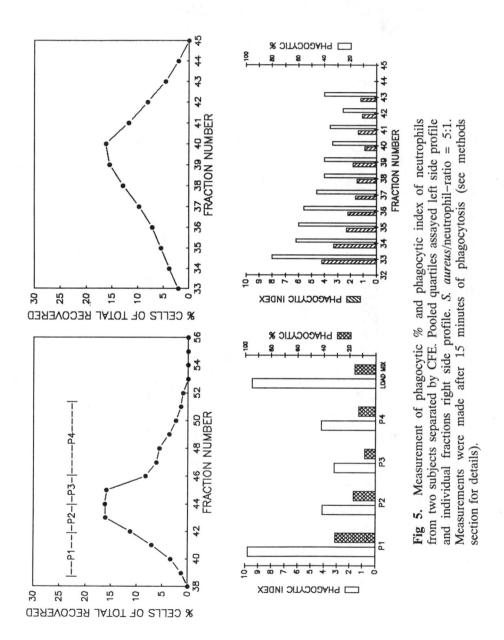

Fig 5. Measurement of phagocytic % and phagocytic index of neutrophils from two subjects separated by CFE. Pooled quartiles assayed left side profile and individual fractions right side profile. *S. aureus*/neutrophil–ratio = 5:1. Measurements were made after 15 minutes of phagocytosis (see methods section for details).

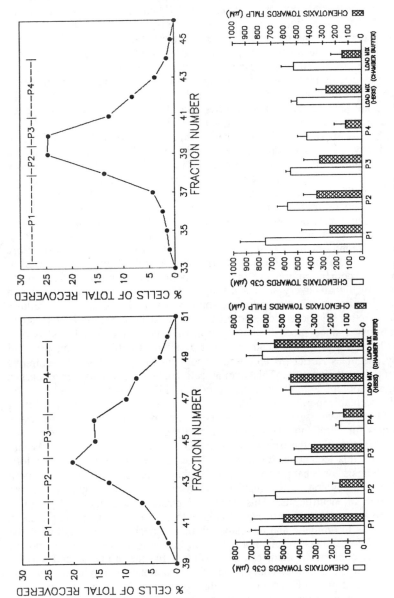

Figure 6. Chemotactic activities of neutrophils from two subjects separated by CFE and pooled (quartiles). Chemotaxis assays used the agarose procedure with FMLP as chemo-attractant.

(i.e. P1 > P2 > P3 > P4) from PAPS/CCD separation profiles. The least electronegative neutrophils thus seem also to be either more metabolically competent or of better receptor status for the activating peptide than the more electronegative neutrophils.

Subpopulation heterogeneity of peripheral blood cells has been studied for many years, but most of the research interest has focused on either red cells in the context of their differences due to *in vivo* senescence or on the separation of T and B lymphocyte subclasses and the study of their immunological characterization and functions. In contrast, the neutrophil has attracted less attention, although it has been long known from clinically based functional studies that, even in normal subjects, some circulating neutrophils respond well to stimuli (chemotaxis, phagocytosis etc.) whilst others appear functionally defective. A knowledge of the normal population heterogeneity of these cells is a necessary prerequisite if we are to understand many disease states affecting neutrophils.

One difficulty in studying isolated neutrophils is their extreme sensitivity to chemical and physical stimuli during the isolation procedure. These stimuli can inadvertently trigger resting or quiescent cells to display motile and secretory functions such that the very basis upon which the separation of subpopulations depends loses its discriminatory advantage. Certainly volume and/or size dependent density gradient separation techniques with sustained hydrostatic pressure and side wall effects during centrifugation may have to be avoided, as also any procedures that involve specific or non-specific binding of activating materials or foreign surfaces to neutrophil membrane surface components (for example, nylon fibre filtration, affinity column fractionation or the attachment of magnetic beads or fluorochromes). Another requirement in cell subpopulation studies is the need to isolate cells with high recovery so that a proper representative sample is obtained and no subpopulation is selectively lost at some stage in the isolation procedure. However, in practice all these idealistic criteria have to be compromised to some extent compromised since speed in the isolation of neutrophils from blood is also desirable with such a short life span cell. The rapid neutrophil isolation protocol described by one of us earlier (8) and used throughout the present studies employs mild differential centrifugation. This avoids cell contact with polymer materials (Dextran, Lymphoprep, Percoll etc.) and produces high yields of cells of good viability, of acceptable purity and with substantially better recoveries than other procedures. Over the course of 2–3 years using this method routinely the many very low values for basal (non-stimulated) capacity of the isolated cells to reduce NBT gives confidence that the isolated neutrophils are essentially quiescent. When, after isolation by this technique, the neutrophils are subjected to either continuous flow electrophoresis or charge-sensitive two-phase aqueous partitioning, both procedures provide evidence of considerable electrokinetic heterogeneity in the circulating pool of neutrophils. This charge-dependent heterogeneity is of a dynamic nature since stimulation of neutrophils with FMLP or their participation in phagocytosis leads to dramatic and reversible shifts in the shape and location of the profile in the CFE chambers (results not presented here). Such profile shifts and the isolated fractions comprising them can be exploited for the study of cell surface changes before, during and after stimulation with inflammatory products. That the distribution profiles from CFE separations (which are generally gaussian) are based upon differences in electrophoretic mobilities has been confirmed by analytical electrophoresis. All cells in the total applied population were electronegative since all cells in complete distribution profile emerged from the CFE separation chamber on the anodal side of the inlet port. Similarly, in the TAPP/CCD separations, the left to right increase in partition ratio correlates well with a left to right increase in cell surface electronegativity as determined by analytical electrophoresis. The single–step partitioning procedure recommended by Walter et al. was used to determine whether or not separated neutrophils in a CCD distribution profile genuinely located according to differences in membrane electrokinetic properties. The left to right partition ratios of neutrophils from 2 subjects increased gradually from 36% to 64% and 37% to 79%, respectively. With the CFE separation procedure, fractions taken from the left or right side of the distributions and re-separated under exactly the same conditions always relocated in the expected positions in the second fractionation profiles. The results support our view that the CCD and CFE distribution curves represent true heterogeneity with respect to the neutrophil surface electrokinetic properties.

Concerning the functional properties of the separated human neutrophils, the results recorded here confirm the earlier findings of Spangenberg & Crawford (6) for pig neutrophils of an inverse relationship between surface electronegativity and phagocytic competence. A similar inverse relationship also exists for chemotactic migration and raises the question of whether the electronegative status of the neutrophil surface in some way controls the disposition or level of macromolecular organization of the cell's cytoskeletal network, which is believed to be the major driving force for these motile events. Both phagocytosis and chemotaxis are cell phenomena that can be inhibited by cytoskeletal poisons such as phalloidin, cytochalasin and colchicine.

In the CFE protocol used for the present studies, the ionic strength conditions for the chamber buffer were chosen to maximize the depth of the electrical double layer or electrokinetic zone surrounding the neutrophils moving in the ionic medium. With low ionic strength buffers this zone is believed to extend from the cell surface almost twice as far (~14 Å) as when buffers of higher conductivity are employed (~8 Å). Under the lower ionic strength conditions chamber buffer flow rates can be increased because particle velocities are higher than at high ionic strengths. Cooling problems are thus reduced.

In conclusion both continuous flow electrophoresis and two phase polymer partitioning as used in these studies have sufficient sensitivity to separate circulating blood neutrophils into fractions substantially based upon differences in cell surface electronegativity. Both techniques are preparative, reasonably easy to handle, and they produce cell fractions suitable for analytical and functional testing. Because of the concern about prior exposure to polymer materials and their possible binding to the cell surface affecting separations in two phase polymer partitioning, a differential centrifugation procedure for the initial cell isolation has been described and recommended. To date we have not been able to provide any evidence that the two separation procedures differ substantially in the cell surface electrokinetic properties they exploit for separation. In both techniques the surface expressed electronegativity, due largely to sialic acid residues on membrane glycoproteins, makes the major contribution to the separations but having fractionated the circulating neutrophil pool on this electrokinetic basis, the heterogeneity is seen to be substantial for functional properties such as phagocytosis and chemotaxis. How much this heterogeneity depends upon surface receptor status for the stimulating agents, or on receptor/ligand linked signal transduction processes in the membrane, or on second messenger generated intracellular responses, is not known at present, but cell fractions prepared by these techniques may provide useful models for studying such processes.

The PAPS/CCD procedure can be carried out in small scale partition devices (2) and although designed mainly for preparative use the CFE apparatus of Bender and Hobein can also be used for small scale studies. Since the chamber outlets of the VAP 22 can be discharged into microtitre wells a two dimensional approach combining a CFE technique with immunological ELISA and FACS analyses is now being used for a more detailed study of the population differences in membranes and other constituents.

Acknowledgments

We thank the Arthritis and Rheumatism Council for financial support (P.E. and N.C.) and also the S.E.R.C. (P.E.) and the Peter Samuel Royal Free Fund (D.F.) for grant support. We also thank Miss Heather Watson for assisting with the preparation of this manuscript.

References

1.   Gallin, J. I.; Durocher, J.; Kaplan, A. *J. Clin. Invest.* **1975,** *55,* 967–973.
2.   Gallin, J. I. *J. Clin. Invest.* **1980,** *65,* 298–306.
3.   Weiss, L. *J. Cell Biol.* **1965,** *26,* 735–739.
4.   Lichtman, M. A.; Weed, R. I. *Blood* **1972,** *39,* 309–316.
5.   Nagura, H.; Asai, J.; Katsumata, Y.; Kojima, K. *Acta Pathol. Japan* **1973,** *23,* 279–290.
6.   Spangenberg, P.; Crawford, N. *J. Cell. Biochem.* **1978,** *34,* 259–268.
7.   Walter, H.; Graham, L. L.; Krob, E. J.; Hill, M. *Biochim. Biophys. Acta* **1980,** *602,* 309–322.
8.   Eggleton, P.; Gargan, P.; Fisher, D. *J. Immunol. Methods* **1989,** *121,* 105–113.
9.   Nelson, R. D.; McCormock, R. T.; Fiegel, V. D. In *Leucocyte Chemotaxis;* Gallin, J. I.; Quie, P. G., Eds.; Raven Press: New York, 1978, pp 25–42.
10. Rook, G. A. W.; Steele, J.; Umar, S.; Dockrell, H. M. *J. Immunol. Methods* **1985,** *82,* 161–167.
11. Sherbert, G. V. *The Biophysical Characterisation of the Cell Surface;* Academic Press: New York, 1978, pp 53–57.
12. Pritchard, D. G.; Halpern, R. M.; Halpern, J. A.; Halpern, B. C.; Smith, C. A. *Biochim. Biophys. Acta* **1975,** *404,* 289–299.

Review Papers

For good reviews of cell separations by partitioning and free flow electrophoresis, see the following:–

Walter, H.; Fisher, D. "Separation and Subfractionation of Selected Mammalian Cell Populations". In *Partitioning in Aqueous Two Phase Systems;* Walter H.; Brooks D.E.; Fisher, D., Eds; Academic Press: New York, 1985, pp 377–414.

Fisher, D. "Separation and Subfractionation of Cell Populations by Phase Partitioning: An Overview". In *Separations Using Aqueous Phase Systems;* Fisher, D.; Sutherland, I. A. Eds.; Plenum Press: New York, 1989, pp 119–125.

Hannig, K.; Heidrich, H. G. *Preparative Electrophoresis of Cells in Free Flow Electrophoresis;* GIT Verlag GMBH: FRG, 1990, Chapter 3.

RECEIVED March 15, 1991

Chapter 14

Separation of Lymphoid Cells Using Combined Countercurrent Elutriation and Continuous-Flow Electrophoresis

Johann Bauer

University of Regensburg, Institute of Pathology, Universitaetsstrasse 31,
D–8400 Regensburg, Germany

A scaled down physical cell isolation procedure including a table-top elutriator and the free flow electrophoresis ACE 710 is described. Three applications to the
study of B-cell maturation are demonstrating that this
system is suitable to isolate small numbers of cells
without alterations of their in-vivo status.

Information about the in-vivo behaviour of cells may be obtained
from in-vitro assays if the cells are isolated without significant
alteration of their in-vivo status. Therefore cell purification
methods were developed which yield highly purified subpopulations of
cells and minimally alter the in-vivo status of the cells. Such
methods include free flow electrophoresis (1), countercurrent
centrifugal elutriation (2), density gradient centrifugation (3) and
flow sorting (4). These methods have several advantages: Cells can
be isolated within a few hours, cells are kept in suspension thus
avoiding contact with surfaces that may provide activation signals,
labeling with antibodies is not required thus avoiding antibodies
that attach to the cell surface and modulate cellular functions
(5,6). To illustrate the utility of combined physical methods of
cell isolation, including electrophoresis, three applications to the
study of B-lymphocyte function are presented in this chapter.

Combinations of Physical Cell Separation Methods

Several laboratories have applied physical cell separation methods
for the purification of cells from peripheral blood, lymphoid organs
and bone marrow (for reviews see 1,7). These cells were preferred,
because it was not too difficult to obtain single cell suspensions.
Recent interest in our laboratory has been focused on the
lymphocyte-monocyte system, especially the purification of B-cells
and the characterization of those B-cells that are in an advanced

state of activation or differentiation (8,9). Two important findings
resulted: None of the methods mentioned above could be used alone in
order to obtain sufficiently pure subpopulations, and physical
stress on the cells must be reduced as much as possible.

Due to the first result we combined several methods to establish
a physical isolation procedure (10). Figure 1 shows the general
procedure which was applied for the purification of cellular
subpopulations from human peripheral blood. It is compared with an
isolation protocol for the C1s component of complement in order to
demonstrate the parallels between enzyme and cell enrichment
procedures that use physical parameters. C1s is precipitated
together with the other euglobulins from human serum. Then it is
enriched according to its negative surface charge and its Stokes
radius. During cell isolation the mononuclear leukocytes (MNL) and
platelets with a density below 1.077 $gcm-3$ are separated from the
rest of the cells in the human peripheral blood by equilibrium
density gradient centrifugation. Then the cells are fractionated
according to their volume into platelets, lymphocytes and monocytes
by counter current centrifugal elutriation (CCE). Finally the CCE
fractions are further purified by free flow electrophoresis
according to their electrophoretic mobility (EM).

Cell Separation Apparatus

Two types of small-volume cell-separation systems were used.

Free Flow Electrophoresis Apparatus ACE 710.   The free flow electrophoresis apparatus ACE 710 (Hirschmann Geraetebau, Unterhaching,
FRG) developed by Hannig (1) was very suitable for fast and gentle
cell isolation. The reduced size of the chamber (4 x 18 cm) makes it
possible to electrophorese cells during only 30 s within the
electric field. Cells with different electrophoretic mobilities (EM)
were only separated by about 1.5 mm, enough separation distance to
fractionate the cells, due to mechanical precision, an efficient,
electronically controlled cooling system, a video system for super-
vising the electrophoresis procedure and a cell injection needle
that focuses the cells into the middle of the buffer film.

Countercurrent Centrifugal Elutriation in a Table-Top Centrifuge.
Another equally important step of the isolation procedure shown
above is countercurrent centrifugal elutriation. In early studies we
used a Beckman Instruments elutriator rotor J-21b. However, the
percentage of cell loss was found to be very high when the rotor was
loaded with fewer than 2×10^8 cells. A new elutriator rotor was
developed to operate well when loaded with about 5×10^7 cells (11),
and its size was adapted to fit a common table-top centrifuge. The
rotor is an 18 cm long steel tube. It contains the separation
chamber and the counterbalancing weight. The separation chamber has
a cylindrical outer shape, its interior was formed in the same shape
as the standard commercially available chamber, but the volume was
reduced to 0.5 ml.

Figure 2 consists of photographs of the new elutriator. In the
lower panel it is partially disassembled. A ring-shaped depression
is milled into the top of the shaft. This ring forms the rotating

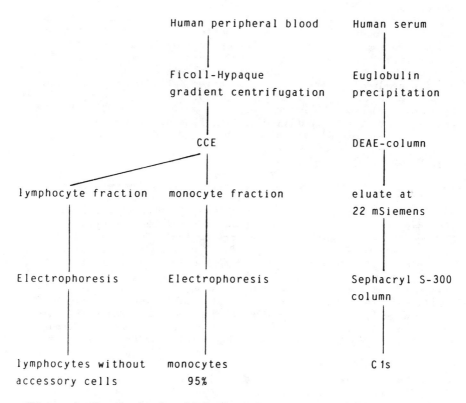

Figure 1: Protocols for isolation of mononuclear leukocytes and of the C1s component of complement illustrating the use of their physical properties. (Adapted from ref. 10).

Figure 2: Photographs of the new elutriator ready for operation (upper panel) or partially disassembled (lower panel) showing the steel-tube rotor. (Adapted from ref. 11).

part of the sealed housing. The stationary part of the sealed
housing is an alloyed delrin block fixed in the centre of the lid
of the centrifuge. During operation both parts are pressed together
by a spring. The rounded piece of the alloyed delrin has two holes.
The central hole leads the incoming buffer to a central tubing
drilled into the shaft. This tubing continues to the bottom of the
separation chamber. The second hole drilled 5 mm to one side of the
first one is the connection tube between the sealed housing and the
outlet hose. The sealed housing is filled by the buffer coming from
the top of the separation chamber.

In the upper panel the system is shown ready for operation. At
the left side of the centrifuge onto which the system is mounted a
piston pump with a 50 ml syringe can be seen. This piston pump
drives about 55 ml medium through the rotor with one stroke so that
counterflow medium sufficient for one separation experiment is
pumped without pulsation at rates of 2.5, 4 and 6 ml/min. The hose
from the syringe to the centrifuge is interrupted by a three-way
valve, which is used to introduce the cells into a spheroidal
chamber fixed between the valve and the centrifuge. At the start of
an experiment the system is filled with medium, the centrifuge is
running and the pump is switched off. After cell injection the
connection between pump and rotor is reestablished, and the cells
are driven from the spheroidal storage chamber into the separation
chamber under constant-pressure conditions. The outlet hose at the
right side is used for collecting the emerging cells.

Application of the Physical Isolation Procedure for B-cell Isolation

By analytical flow cytometry and by plaque forming assays it was
previously demonstrated that three kinds of B-cells are found in
human peripheral blood (12-15): Resting B-cells that can be
triggered by antigens or mitogens to mature to antibody-producing
plasma cells, in-vivo preactivated B-cells that have already
received their signals but still need growth factors such as IL-4,
5, 6 or IL-1 and IL-2, and finally B-cells in a final state of
maturation that secrete antibodies while they are circulating.
Using our isolation system we tried to separate B-cells in the
different stages from each other on a preparative scale and to
enrich them as far as possible.

Figure 3 shows that it was possible to isolate resting in-vivo
preactivated and mature B-cells from the peripheral blood. The solid
line with stars indicates antibody production by isolated resting
B-cells. These cells had to be stimulated with pokeweed mitogen, and
they started to secrete antibodies on day 5 after the beginning of
incubation. The solid line with open circles and the dashed line
show the antibody production of in-vivo preactivated and mature
B-cells. Both of these cell types produced antibodies even in the
absence of any defined antigen or mitogen; however, they differed in
their onset of antibody release. The mature B-cells started secreting antibodies on the first day of incubation, and the preactivated
cells on day 3. The populations were further analysed and
characterized (9).

It was found that the in-vivo preactivated B-cells that started antibody secretion on day 3 of incubation had an EM of 0.9×10^{-4} (cm^2 V^{-1} s^{-1}), similar to that of the resting cells. Their antibody production capacity could be enhanced by IL-2 and other growth factors, and they produced mainly IgM. The mature B-cells, which secreted antibodies on day 1 of incubation had an increased electrophoretic mobility similar to that of normal T-cells (1.1×10^{-4} (cm^2 V^{-1} s^{-1})). They produced optimal quantities of antibodies irrespective of whether or not IL-2 or growth factors were added, and they secreted mainly IgG.

In-vitro Assays of B-cells Fractionated by CCE and Free Flow Electrophoresis. The separation experiments showed that the physical isolation methods provide a source of B-cells in their native state of activation or differentiation. These B-cells could be used for in-vitro analysis of B-cell functions (Table I). Three functions were investigated.

The mechanism of the increase of the electrophoretic mobility that occurs when B-cells progress from the preactivated to the mature state of differentiation was studied.

Isolated B-cell populations were used to obtain information about the action of a synthetic glycolipid that is considered to enhance antibody production.

Finally, it was determined whether or not B-cells that produce detectable quantities of antibodies spontaneously can be isolated from patients suffering from a locally confined tumor.

Clinical Study. A clinical study was performed in collaboration with Dr. U. Krinner, University of Regensburg, who provided 60 ml of whole blood drawn from healthy donors and from cancer patients with a squameous epithelial tumor within the oral cavity. The mononuclear leukocytes (MNLs) were isolated from the blood samples and fractionated according to their volume by CCE. The cells from different fractions were incubated for 7 days in the absence of a growth factor or a mitogen, and the level of antibodies released into the supernatant medium were determined. To date it has been found that most healthy blood donors possess cell fractions whose B-cells spontaneously produce more than 600 ng/ml Ig in 7 days. None of 9 tumor patients investigated so far possessed cell fractions that produced more than 600 ng/ml Ig in 7 days in-vitro. The study will be continued until statistically significant results are obtained. Nevertheless, the preliminary results show that the improved, scaled-down physical isolation techniques can be used for clinical studies.

Pharmacological Study. For performing a pharmacological study Dr. K. G. Stuenkel from the BAYER research center in Wuppertal, FRG kindly provided the synthetic glycolipid BAY R 1005, which is reported to have immunoenhancing activity (16,17). BAY R 1005 did not have mitogenic activity, because it did not induce resting B-cells to multiply and produce antibodies. Mature plasmacytoid B-cells with high EM secreted equal quantities of antibodies in the presence and in the absence of the compound. However, the in-vivo preactivated B-lymphocytes were triggered by BAY R 1005 to increase

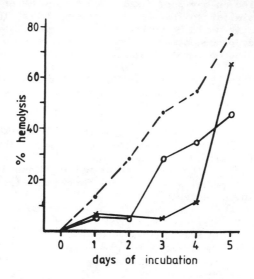

Figure 3: Antibody production by three classes of B-lymphocytes.
Hemolysis of protein A-coated sheep red blood cells caused by
200 μl culture supernatant and guinea pig complement was used as
immunoglobulin assay. Antibodies were produced by small resting
B-lymphocytes cultured in the presence of T-lymphocytes,
monocytes and pokeweed mitogen (solid line with stars), by large
B-lymphocytes with low EM (solid line with open circles) or high
EM (dashed line). (Adapted from ref. 9).

Table I: Studies Performed with B-cell Subpopulations Isolated by
Physical Methods

	Questions	Results
Basic research	Does the EM of B-cells increase during final maturation?	Yes
Pharmacological study	Does Bay R 1005 have an effect on human B-cell function?	Yes
Clinical study	Do locally confined tumors suppress antibody production?	(Yes)

their antibody production. Furthermore BAY R 1005 triggered
monocytes to secrete a factor into the supernatant medium, which
supported antibody production during a mixed lymphocyte reaction.
The results demonstrated that cells isolated by these physical
methods are suitable for testing new pharmaceutical compounds
(manuscript in preparation).

Basic Research. In the third study the mechanism of the EM
increase during B-cell maturation was investigated. It had been
shown that mature B-cells with high EM are further differentiated
than the in-vivo preactivated B-cells with low EM (9) (Figure 3).
Thus the first idea was that a change of the EM occurs when the
preactivated B-cells further mature to antibody secreting cells.
When preactivated B-cells were incubated in-vitro for 3 or 5 or 7
days, the cells produced antibodies; however, their EM did not
increase. So the question remained open whether there are a few
B-cells in the human circulation with high EM, which are only
detectable after activation, or whether B-cells are only able to
change their EM during final maturation in-vivo. In order to address
this question we used the mouse myeloma cell line Ag8 as a model for
final B-cell maturation (18).

Figure 4 shows two different EM distribution curves of Ag8
cells. The cells indicated by the dashed curve were harvested from a
culture with high cell density. The curve resembles an EM
distribution curve of normal human or mouse B-cells. The solid curve
was obtained from an analysis of Ag8 cells that had been cultured at
low cell density. A lot of cells (up to 50%) have higher electrophoretic mobility. The cells were separated on a preparative scale
into three fractions with EM between 0.92 and 0.99, between 1.0 and
1.06, and between 1.07 and 1.13. Analytical reelectrophoresis proved
that the cells in the different electrophoretic fractions truly had
different EMs and that the alteration of the EM distribution curve
was not a procedure-dependent artifact. In order to find the reason
why myeloma cells had increased their EM during cultivation at low
cell density, we characterized the cells found in each electrophoresis fraction. The cells with enhanced EM showed lower
proliferation activity and lower clone forming activity but higher
alkaline phosphatase activity, which is considered a differentiation
marker of very mature B-cells (19). Evidently some of the myeloma
cells spontaneously completed a step of differentiation toward their
final state of maturation. The results obtained in the mouse myeloma
cell experiments suggested that B-cells undergo an increase in
negative surface charge density during final maturation (18). At the
moment we investigate the expression of the alkaline phosphatase on
human B-cells.

Conclusion

These types of experiments performed with isolated B-cells
demonstrated that B-cells isolated by appropriately adjusted
physical methods are suitable for experiments that involve in-vitro
assays. They could be used for basic research, clinical study and
pharmacological study.

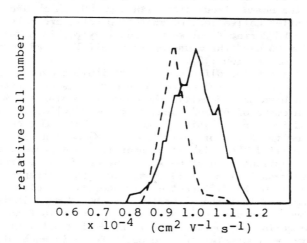

Figure 4: EM distribution curves of myeloma cells harvested from cultures with high cell density (dashed line) or low cell density (solid line).

This type of experience obtained from lymphocyte fractionation can be applied, for example, to the isolation and purification of tumor cells and other cells less readily available for study in suspension.

Literature Cited

(1) Hannig, K., Electrophoresis 1982, 3, 235-243.
(2) Sanderson, R.J., Shepperdson, F.T., Vatler, A.E. and Talmage, D.W., J. Immunol. 1977, 118, 1409-1414.
(3) Boyum, A., Scand. J. Clin. Invest. 1968, 21, Suppl.97, 31-39.
(4) Melamed, M.R., Mullaney, P.F. and Mendelsohn M.L., Flow Cytometry and Sorting, Wiley New York, 1979.
(5) Van Wauwe, J.P., de May, J.R. and Goosens, J.G., J. Immunol. 1980, 124, 2708-2713.
(6) Chiorazzi, N., Fu, S.M. and Kunkel, H.G., Clin. Immunol. Immunopathol. 1980, 15, 380-391.
(7) Bauer, J., J. Chromatogr. 1987, 418, 359-383.
(8) Bauer, J. and Hannig, K., Electrophoresis 1986, 7, 367-371.
(9) Bauer, J., Kachel, V. and Hannig, K., Cell. Immunol. 1988, 111, 354-364.
(10) Bauer, J. and Hannig, K., Electrophoresis `86; M.J. Dunn: VCH, Weinheim, 1986, pp. 13-24.
(11) Bauer, J. and Hannig, K., J. Immunol. Methods 1988, 112, 213-218.
(12) Kikutani, H., Kimuro, R., Nakamura, H., Sato, R., Muraguchi, A., Kawamura, N., Hardy, R.R. and Kishimoto, T., J. Immunol. 1986, 136, 4019-4026.
(13) Anderson, K.C., Roach, J.A., Daley, J.F., Schlossman, S.F. and Nadler, L.M., J. Immunol. 1986, 136, 3012-3017.
(14) Kehrl, J.H., Muraguchi, A., Butler, J.L., Falkoff, R.J.M. and Fauci, A.S., Immunol. Rev. 1984, 78, 75-98.
(15) Thomson, P.D. and Harris, N.S., J. Immunol. 1977, 118, 1480-1485.
(16) Lockhoff, O., Hayauchi, Y., Stadler, P., Stuenkel, K.G., Streissle, G., Paessens, A., Klimetzek, V., Zeiler, H.J., Metzger, K.G., Kroll, H.P., Brunner, H. and Schaller, K., German Patent, No. DE 35219994, 1987/1985.
(17) Stuenkel, K.G., Lockhoff, O., Streissle, G., Klimetzek, V., Lockhoff, O. and Schlumberger, H.D., Cellular Basis of Immune Modulation. Progress in Leukocyte Biology.; Kaplan, J.G., Green, D.R. and Bleakley, R.C.: Alan R. Liss. Inc., New York, 1989, Vol. 9, pp. 575-579.
(18) Bauer, J. and Kachel, V., Immunol. Invest. 1990, 19, 57-68.
(19) Burg, D.L. and Feldbush T.L., J. Immunol. 1989, 142, 381-387.

RECEIVED March 15, 1991

Chapter 15

Comparison of Methods of Preparative Cell Electrophoresis

Paul Todd

Center for Chemical Technology, National Institute of Standards and Technology, 325 Broadway 831.02, Boulder, CO 80303–3328

In the electrophoretic separation of macromolecules, resolution, purity and yield establish the merits of a separation method. In the separation of living cells by electrophoresis, additional significant criteria include viability, capacity, convenience and cost, as the method must also be acceptable to cell biologists, nondestructive to living cells and in free solution. These demands lead to the following quantitative characteristics of methods of cell electrophoresis: separation (resolution), purity, capacity, product viability, convenience and capital cost. Thirteen different methods of electrophoretic cell separation were compared on the basis of these six criteria using a relative scale of 0 to 4. The methods compared were free zone electrophoresis, density gradient electrophoresis, ascending vertical electrophoresis, re–orienting density gradient electrophoresis, free–flow electrophoresis, agarose sol electrophoresis, low–gravity electrophoresis, continuous flow low–gravity electrophoresis, rotating annular electrophoresis, density–gradient isoelectric focusing, free flow isoelectric focusing, stable flow free boundary electrophoresis, and continuous magnetic belt electrophoresis. In each case the sum of the six scores provides an overall "figure of merit"; however, no method rates a perfect score of 24. While quantitative trade-offs, such as resolution for capacity and vice versa, impose technical limitations on each method, lack of convenience is one of the most significant factors in the failure of cell electrophoresis to become a widely popular method.

Cell electrophoresis, like the electrophoresis of macromolecules, is a high–resolution separation method that is difficult to scale up.

Subpopulations of cells for which no affinity ligand has been developed and for which there is no distinct size or density range are often separable on the basis of their electrophoretic mobility, which may in turn be related to their function, whether fortuitously or otherwise.

Electrophoresis

Electrophoresis is the motion of particles (molecules, small particles and whole biological cells) in an electric field and is one of several electrokinetic transport processes. The velocity of a particle per unit applied electric field is its electrophoretic mobility, μ; this is a characteristic of individual particles and can be used as a basis of separation and purification. This separation method is a <u>rate</u> (or transport) process.

The four principal electrokinetic processes of interest are electrophoresis (motion of a particle in an electric field), streaming potential (the creation of a potential by fluid flow), sedimentation potential (the creation of a potential by particle motion), and electroosmosis (the induction of flow at a charged surface by an electric field). These phenomena always occur, and their relative magnitudes determine the practicality of an electrophoretic separation or an electrophoretic measurement. It is generally desirable, for example, to minimize motion due to electroosmosis in practical applications. A brief discussion of the general electrokinetic relationships follows.

The surface charge density of suspended particles prevents their coagulation and leads to stability of lyophobic colloids. This stability determines the successes of paints and coatings, pulp and paper, sewage and fermentation, and numerous other materials and processes. The surface charge also leads to motion when such particles are suspended in an electric field. The particle surface has an electrokinetic ("zeta") potential, ζ, proportional (in a given ionic environment) to σ_e, its surface charge density — a few mV at the hydrodynamic surface of stable, nonconducting particles, including biological cells, in aqueous suspension. If the solution has absolute dielectric constant ϵ, electrophoretic velocity is

$$v = \frac{2\zeta e}{3\eta} E \tag{1}$$

for small particles, such as molecules, whose radius of curvature is similar to that of a dissolved ion (Debye–Hückel particles), and

$$v = \frac{\zeta e}{\eta} E \tag{2}$$

for large ("von Smoluchowski") particles, such as cells and organelles. These relationships are expressed in rationalized MKS (SI) units, as clarified, for example, by Hunter (<u>1</u>), where η is the viscosity of the bulk medium, and ϵ is the dielectric constant. At typical ionic strengths (0.01 to 0.2 equiv./L) particles in the nanometer size range usually have mobilities (v/E) different from those specified by equations (1) and (2) (2).

Preparative Electrophoresis

Analytical electrophoresis in a gel matrix is very popular, because convection is suppressed; however, high sample loads cannot be used owing to the limited volume of gel that can be cooled sufficiently to provide a uniform electric field, and cells do not migrate in gels. Capillary zone electrophoresis, a powerful, high-resolution analytical tool (3), depends on processes at the micrometer scale and is not applicable to preparative cell electrophoresis. Therefore preparative electrophoresis must be performed in free fluid, so great care must be taken to prevent thermal and solutal convection and cell sedimentation. The two most frequently used free-fluid methods are zone electrophoresis in a density gradient and free-flow (or continuous flow) electrophoresis (FFE or CFE). Other, less popular, methods will not be discussed in detail. These are described in partial reviews by Ivory (4) and by Mosher et al. (5).

Physical Constraints Imposed by Living Cells

Buffers. Typical categories of buffers used with each of the methods possess a wide range of ionic strengths, partly to avoid the application of high currents in preparative separations, which require low-conductivity buffers to avoid convective mixing due to Joule heating. At very low ionic strength, phosphate buffers are found harmful to cells, and this is the principal reason that triethanolamine was introduced as a general carrier buffer for free-flow electrophoresis by Zeiller and Hannig (6). When cells are present neutral solutes such as sucrose or mannitol must be added for osmotic balance. These buffers have undergone testing in analytical and preparative cell electrophoresis.

Temperature. Temperature control is generally imposed on all free electrophoresis systems as a means of suppressing thermal convection and avoiding viscosity gradients. Biological separands impose additional constraints on thermoregulation of electrophoretic separators. With a few notable exceptions, living cells do not tolerate temperatures above 40°C. Additionally, mammalian cells are damaged by being allowed to metabolize in non-nutrient medium, so it is necessary to reduce metabolic rate by reducing temperature, typically to 4 to 10°C. Proteins are more thermotolerant, but at 37°C the enzymes of homeothermic organisms and many bacteria (including E. coli) are active, and unwanted hydrolysis reactions occur in samples containing mixtures of proteins.

Electric field. It is possible to move membrane proteins in cells immobilized in an electric field (7) and, in sufficiently strong fields (> 3 kV/cm) to produce dielectric breakdown of the plasma membrane in a few μs, resulting in a potential for cell-cell fusion or pores through which macromolecules can readily pass (8). The strength of such fields, 1 to 15 kV/cm, considerably exceeds the range of field strength used for electrophoresis.

Role of cell size and density. Over a wide range of radii the mobility
of a given cell type is independent of cell size, as predicted by
electrokinetic theory (Equation (2), above). Isolated electrophoretic
fractions collected after density gradient electrophoresis have been
analyzed individually with a particle size distribution analyzer, and
no consistent relationship between size and electrophoretic fraction
was found (9). This finding implies that, although the gravity and
electrokinetic vectors are antiparallel (or parallel) in density
gradient electrophoresis, size plays very little role in determining
the electrophoretic migration of cultured cells. As earlier chapters
have implied, density plays an important role in cell sedimentation,
and when $\rho - \rho_o$ is excessive (>0.03 g/cm^3) free electrophoresis in a
density gradient is dominated by sedimentation, while separation by
free-flow electrophoresis is less so (10, 11).

Droplet Sedimentation

The diffusion coefficients, D, of small molecules are in the range 10^{-6}
-10^{-5} cm^2/s, of macromolecules, $10^{-7}-10^{-6}$; and of whole cells and
particles, $10^{-12}-10^{-9}$. If a small zone, or droplet, of radius R
contains n particles of radius a inside, whose diffusivity is much less
than that of solutes outside, then rapid diffusion of solutes in and
slow diffusion of particles out of the droplet (with conservation of
mass) leads to a locally increased density of the droplet:

$$\rho_D = \rho_o + n\frac{a^3}{R^3}(\rho_D - \rho_o) . \qquad (3)$$

If $\rho_D > \rho_o$ then the droplet falls down; if $\rho_D < \rho_o$ it is buoyed upward
(12). Droplet sedimentation (or buoyancy) is a special case of
convection, and its occurrence is governed by the following stability
rule

$$D(d\rho_o/dz) + D_S(d\rho/dz) > 0 \qquad (4)$$

in which D and D_S are diffusivities of the cell and solute,
respectively. Typically dρ/dz will be somewhere negative, and dρ_o/dz
= 0 except in density gradient electrophoresis. In column
electrophoresis droplet sedimentation cannot be prevented except at low
cell densities (13), so a suitable way to scale column electrophoresis
is to increase column cross-sectional area (14). Under conditions of
droplet sedimentation particles still behave individually unless the
ionic environment also permits aggregation (15, 16). In the case of
erythrocytes there is sufficient electrostatic repulsion among cells
to permit the maintenance of stable dispersions up to at least 3 x 10^8
cells/mL (17,18). Droplet sedimentation is an avoidable initial
condition in zone processes, but it can be created during processing
when separands become highly concentrated as in isoelectric focusing.
The absence of droplet sedimentation in low gravity is one of the most
significant attractions of low-gravity separation science.

Evaluation of Separation Methods:

Resolution

Two definitions of resolution are used in this review (see Chapter 1). One follows the paradigm of multi-stage extraction or adsorption and identifies the practical limit of the total number of fractions that can be collected from a given separation device, NSU ("number of separation units"), which is usually specified in equipment design. The inverse view is also useful. In sedimentation, in which resolution depends on the standard deviation of the distance sedimented and electrophoresis, in which it depends on the standard deviation of mobility, and in flow sorting, where it depends on the standard deviation of optical fluorescence, for example — all cases in which the moments of the appropriate measurement are known — the coefficient of variation ("CV") or "relative standard deviation," in any case the ratio of standard deviation to the mean is used. CV is not a simple reciprocal of NSU, as the measured electrophoretic standard deviation accounts for physical distortion of separand zones during migration and collection, intrinsic biological heterogeneity and counting statistics in addition to the finite size of collected fractions. So, as a rule,

$$CV > 1/NSU. \tag{5}$$

Thus both values will be included, as they express both the geometrical and practical resolution of each instrument.

Purity

In the case of cell separation purity is defined as the proportion of cells in a separated fraction that are of the desired type. In multifraction "peaks," single fractions can be identified with greater than 99% purity; however, individual fractions may have low volume (and hence low cell numbers), and a whole "peak," when pooled, may have much lower purity. Purity and yield are thus always related. The overlap of peaks is more severe in some separation systems than in others, and even narrow peaks can be heavily contaminated with cells from adjacent fractions. Therefore it is meaningful to use the maximum reported purity as an indicator for each method.

Viability

It is usually possible to contrive separation conditions that do not kill living cells. In some cases, ingenuity is required to minimize shear forces, eliminate toxic chemicals (including certain affinity ligands), incorporate physiologically acceptable buffer ions, maintain osmolarity, control the temperature, etc. In nearly all cases, abnormal salt concentrations are required, so the system that can use the highest salt concentration will produce the highest viability. This, in turn, is the system with the most effective heat rejection. It is likely, although not argued in this review, that minimum Rayleigh number, Ra, will result in maximum viability.

Capacity

Capacity (also known as "throughput") is measured as mass of feedstock processed per hour or mass of product separated per hour, depending on the objective. This figure applies whether processing is continuous or batchwise. In the case of cell separation, especially by electrophoresis, typical units of capacity would be millions of cells per hour. Because, under most electrophoresis conditions, the maximum feed concentration is 10^7 cells/mL (depending slightly on cell size and density), then capacity is directly related to the volume of feed that can be processed per hour.

Convenience

In the labor-intensive arena of biomedical research convenience may come in one of two forms: a simple process that requires very little engineering skill or physical manipulation or a complex process that has been highly automated. Biomedical researchers do not wish to dedicate excessive amounts of manpower to separation process development or maintenance. In cell electrophoresis the formation of a density gradient or the mixing of a multitude of buffers may be considered an inconvenience; the need to travel to a unique facility is also a deterrent to multiple experiments.

Cost

The cost of a separation process includes capital equipment (apparatus), reagents and labor. The economics of investing in separation technology are highly dependent upon goals. A frequently repeated process, for example, is better done by an automated system and is capital-intensive. A rarely performed, intellectually demanding procedure would be labor-intensive. And a process that requires large quantities of expensive affinity reagents (such as antibodies) would not be used to separate large quantities of separand without provisions for recycling ligands or reducing their cost. Thus the "cost" of a cell electrophoresis method is the sum of capital, labor and supplies. Over the full range of methods considered total cost ranges over 4 orders of magnitude. In this review only capital cost is evaluated.

Methods of Preparative Cell Electrophoresis

There are two very broad categories of cell electrophoresis methods. The following discussion divides all methods into static column methods, which are discussed first, and flowing methods. Static column methods are batch processes while flow methods are continuous. Each method is introduced by a quantitative description and then evaluated according to the 6 criteria.

1. Free Zone Electrophoresis (FZE)

Horizontal column electrophoresis has been performed in rotating tubes of 1-3 mm inner diameter (19). The small diameter minimizes convection distance and facilitates heat rejection while rotation counteracts all three gravity-dependent processes: convection, zone sedimentation and particle sedimentation. The concept is illustrated in Figure 1, which shows that the total vertical velocity vector oscillates as a sample zone moves in the electric field, so that spiral motion results with a radius vector that can be derived from equations of motion in which the sample zone is treated as a solid particle. In actuality, the sample zone is more like a sedimenting droplet, which can be treated as a particle with density ρ_D (see "Droplet Sedimentation," equation (3), above). Due to gravity and centrifugal acceleration, the center of the spiral in Figure 1 (coordinates k,l) is not the center of the tube, and the vertical circle, in x and y, described by the sample zone is

$$(x-k)^2+(y-l)^2=r^2\exp(2\gamma t) \tag{6}$$

in which a zone is stabilized in suspension when k^2+l^2 and γ, the rate constant for centrifugal motion, are minimized. Hjertén solved the equations of motion for these values and found that if the Hjertén number, recently defined ($\underline{20}$) as

$$Hj \equiv \gamma\tau, \tag{7}$$

is < 1.0, then acceptable conditions for stability of the sample zone exist. In a typical separation τ, the residence time is between 1000 and 10000 sec, and γ is between 10^{-5} and 10^{-3} s^{-1}.

In an open system, in which fluid flow in the direction of the applied electric field is possible, the net negative charge on the tube wall results in plug—type electroosmotic flow (EEO) that enhances total fluid transport according to the Helmholtz relationship ($\underline{21}$) integrated over the double layer at the chamber wall

$$v=\frac{\zeta e}{\eta}E. \tag{8}$$

In a closed system, such as is used in almost all practical cases, flow at the walls must be balanced by return flow in the opposite direction along the center of the cylindrical tube, so that the flow profile is a parabola ($\underline{22}$, $\underline{23}$, $\underline{24}$)

$$v=\frac{\zeta e}{\eta}E(r^2/R^2-1/2) \tag{9}$$

Thus, if the electroosmotic mobility μ_{eo} (coefficient of E in equation (8)) is high (of the order 10^{-4} cm^2/V-s), then sample zones will be distorted into parabolas as shown in the simulation of Figure 2. An additive parabolic pattern is superimposed if there is a radial temperature gradient that causes a viscosity gradient ($\underline{19}$). High temperature in the center of the chamber results in low viscosity and higher velocity due to both EEO return flow and faster migration of separands. Increased temperature also increases conductivity. Excessive power input under typical operating conditions can result in

THE FREE-ZONE ELECTROPHORESIS PRINCIPLE

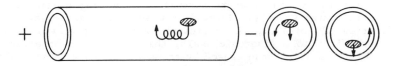

Figure 1.   Free zone electrophoresis.   Trajectory followed by a
particle in FZE in a rotating tube.   In the presence of gravity
the center of the spiral is below the center of rotation (19).
(Reproduced with permission from reference 20. Copyright 1990 American
Institute of Aeronautics and Astronautics.)

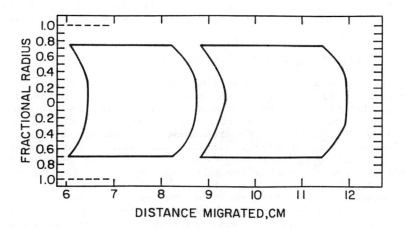

Figure 2.   Contour lines of sample bands predicted in simulation
of human (right) and rabbit (left) erythrocyte migration, from
left to right, in a cylindrical column in low gravity, using
simulation method of Vanderhoff and Micale (1979; Micale et al,
1976).   Assumptions were τ = 1.0 h, μ_{eo} = -0.06, μ(human) = -
2.05, μ(rabbit) = -1.05 x 10^{-4} cm^2/V-s, E = 18.6 V/cm, sample
width = 0.75 x chamber diameter.   Courtesy of F. J. Micale.
(Reproduced with permission from reference 20. Copyright 1990 American
Institute of Aeronautics and Astronautics.)

25% higher separand velocity in the center of the tube even in the absence of EEO (19). In small—bore tubes with closed ends, scrupulous care must be taken to suppress EEO by using a nearly neutral, high—viscosity coating (19), and rapid heat rejection and/or low current density is necessary to minimize radial thermal gradients.

All of the above problems were considered in the engineering of a free zone electrophoresis system with a 20 cm long x 3 mm diameter rotating tube with no EEO (19, 25). A practical minimum collection volume of 0.05 mL from a total of 1.5 mL gives NSU = 30, and the results of optical scanning of the electrophoresis tube during a run shown in Figure 3 indicates CV = 5%. Owing to the shortness of a single batch purification (20 min) and the efficient heat transfer, which allows experiments in 0.15 M (isotonic) salt, this system has no effects on cell viability beyond those caused by spending the same amount of time in suspension, except possibly in those cases where aggregation or other cell—cell interactions occur (25, 15). The electric field is orthogonal to the gravity vector, and rotation compensates for convection, so droplet sedimentation can occur freely, and the maximum cell load is estimated at around 10^8 cells in a starting zone of 0.1 mL (very densely packed). A 30—min processing time gives capacity = 2×10^8 cells/h. The main sources of inconvenience are the preparation of the electrophoresis tube with low EEO and the meticulous task of loading the sample. Only 1 or 2 such units are in operation in the world. The cost of instrument production, one—at—a—time, should be $30000 − $50000.

2. Density Gradient Electrophoresis (DGE)

Density gradient electrophoresis has been used for protein and virus separation (26) and has been shown to separate cells on the basis of electrophoretic mobility as measured by microscopic electrophoresis (27, 28). Methods and applications of this technique have been reviewed (29). Downward electrophoresis in a commercially available device utilizing a Ficoll gradient was first used to separate immunological cell types (30, 31, 32). Similar gradient—buffer conditions are employed in the Boltz—Todd apparatus (13, 23) in which electrophoresis of cells and other separands is usually performed in the upward direction (Figure 4). The use of a density gradient counteracts (within calculable limits) all three effects of gravity: particle sedimentation, convection and zone, or droplet, sedimentation. The migration rates of cells in a density gradient are, however, quantitatively different from those observed under typical analytical electrophoresis conditions (6, 24), because the following physical properties are not depend on position in the density gradient: (1) The sedimentation component of the velocity vector changes as the density of the suspending fluid decreases. (2) The viscosity of the suspending Ficoll solution decreases rapidly with increasing height (33). (3) The conductivity of the suspending electrolyte increases, consistent with the viscosity reduction. (4) Carbohydrate polymers, including Ficoll, typically used as solutes to form density gradients increase the zeta potential of cells (34, 35). (5) Cells applied to the density gradient in high concentrations undergo droplet sedimentation (12, 13) thereby increasing the magnitude of the sedimentation component of the

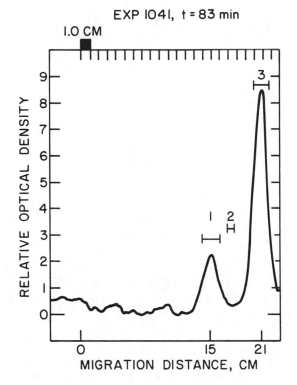

Figure 3. Electrophoretic separation of chicken and rabbit erythrocytes (fixed with glutaraldehyde) by FZE. Sampling regions 1, 2 and 3 were used to determine purity, and data were used to determine CV.

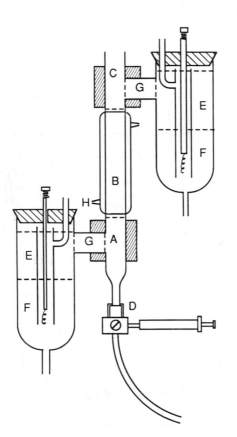

Figure 4.   Diagram of a DGE column labeled to indicate the location of various solutions at the start: A. Top buffer, B. Migration zone with or without Ficoll gradient containing gelling material, C. Sample to be separated, D. Dense solution to support solutions in the column, E. Top solution which prevents electrode solutions from entering the column, F. Electrode solution in which is immersed a bright platinum wire electrode, G. Polyacrylamide gel plug to support hydrostatic pressure of the column, H. Thermostated jacket. Typically B contained a gradient of 1.7 – 6.2% Ficoll and an inverse gradient of 6.5 – 5.8% sucrose with or without agarose, D contained 15% Ficoll, and F was saturated NaCl (13).

migration velocity in equation (10), below, by replacing it with equation (3), above.

All of these variables have been studied (10, 33, 35, 36, 37). In addition cell population heterogeneity with respect to both density and size contributes to the distribution of migration rates (10).

The instantaneous migration velocity of a particle (assumed negatively charged) undergoing upward electrophoresis in a vertical density gradient may be described by

$$dx/dt = \mu(x) E(x) - 2a^2 g(\rho - \rho_o(x))/9\eta(x) \tag{10}$$

where, in typical situations of interest, inertial and diffusion terms are neglected. In equation (12), x is defined as vertical distance from the starting zone, μ is the anodic electrophoretic mobility, E is the electric field strength, usually determined from

$$E(x) = I/(Ak(x)) \tag{11}$$

in which I is the constant applied current, k is the conductivity of the gradient solution, and A is the cross-sectional area of the gradient column. The second term of equation (10) represents the Navier-Stokes sedimentation velocity for spheres of density ρ and radius a under the influence of the gravitational acceleration g. The x-dependent variables are explicitly indicated in equations (10) and (11). Standard electrophoretic mobilities, μ_s, are commonly expressed at the viscosity of pure water at 25°C, η_s. Thus

$$\mu(x) = \mu_s \eta_s / \eta(x). \tag{12}$$

Explicit expressions have been established for the x-dependent variables to allow integration of equation (10) to yield the distance of migration up the column as a function of time of application of the electric field. These empirical relationships consist of treating density and mobility as linear functions of C(x), the polymer (or density gradient solute) concentration, conductivity as a quadratic function of C(x), and viscosity as an exponential function.

Substitution of these functions into equation (10) gives an explicit relationship between velocity and migration distance. The time t(x) to migrate a distance x may be found by numerical integration or analytically in terms of exponential integrals (37):

$$t(x) = \int \frac{Q(x) \exp(\alpha x)}{L(x)} dx \tag{13}$$

Q(x) is a quadratic function derived from the dependence of measured conductivity on C(x), exp(αx) is derived from the measured dependence of viscosity on C(x), and L(x) is a linear function of C(x) derived from the measured dependence of electrophoretic mobility on C(x). Equation (13) can predict experimentally observed migration distance-vs.-time plots without using fitted parameters (10, 35). In cases of a narrow cell density distribution the gradient may be tailored so that the cells are close to neutral buoyancy over an appreciable distance of the column.

Cell separation by electrophoresis in a Ficoll density gradient is typically accomplished in a 2.2 x 7.0 cm water-jacketed glass column using an isotonic gradient from 1.7 to 6.2% (m/v) Ficoll. Because no fluid flows are present during the separation process, sample removal is the only source of significant sample band dispersion. When sample loads are too high, additional band broadening is caused by droplet sedimentation (13), a gravity-dependent phenomenon that adversely affects all free electrophoresis processes in which the sample is applied as a zone (12).

In a typical column of 30 mL volume, 0.05-mL fractions have been collected, giving NSU = 600. From Figure 5 (13) it can be estimated that CV = 5%, which seems to be the lower limit for any preparative cell electrophoresis method. Purity in a model cell separation has been reported at 99.8% (28). Viability decrements to 85% have been observed during run times and/or holding times of the order of a few hours. As mentioned above, capacity is compromised by droplet sedimentation, in a 2.2 cm diameter column it is limited to 1×10^7 nucleated cells in a 1 mL volume; in a 2-h separation the capacity is 5×10^6 cells/h. The principal inconvenience of this method is the forming of the density gradient; however, this is a routine procedure in some professions, especially including experimental biochemistry. Access to apparatus depends on the skill of the investigator in glassware construction and/or the accessibility of glassblowers and machinists. Including a power supply and gradient maker, a system can be acquired for around $2000.

3. Ascending Vertical Electrophoresis (AVE)

A limited number of separation problems can be solved by combining cell sedimentation and electrophoretic "levitation" at low (<10 V/cm) field without a density gradient (38, 39). Suitable apparatus does not differ from the water-jacketed vertical cylinder (Figure 4), and the migration equation is a simplified version of Equation (10), in which viscosity, conductivity, and mobility are simple constants.

Electrophoretic resolution is difficult to quantify in this system, because two intrinsic physical variables are used to drive separation: μ_e and a. Resolution data have not been reported, but NSU can be estimated from a typical 6.6 mL column volume, which gives NSU = 130, assuming a minimum fraction size of 0.05 mL. Purities around 91-100% have been reported on the basis that a population of non-antigen-producing lymphocytes could be separated from a lymphocyte mixture. Model-system separations were not performed. On the basis of the multi-parameter separation it can be assumed that purity is lower than in other column methods. Presumably the droplet-sedimentation limit applies to this method, and it may be more severe, owing to the absence of solutes that increase the buffer density. Proper processing practice would limit this separation to about 2×10^6 cells/h, and 7×10^6 is possible in a D_2O gradient (39). Convenience of this method

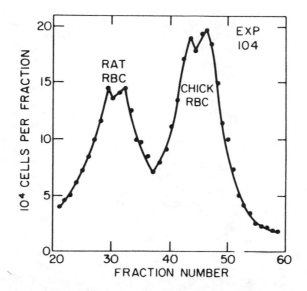

Figure 5. Density gradient electrophoresis. Fractionation
profiles of mixed fixed rat and chicken erythrocytes after
electrophoresis at low cell density. These cell types differ in
mobility by 0.12 x 10^{-4} cm²/V–s, and the two peaks shown are
separated by 10 fractions of 0.25 mL volume each. The
flattening and apparent splitting of the peaks is considered to
be caused by droplet sedimentation. (Reproduced with permission from
reference 13. Copyright 1979 Elsevier/North-Holland Biomedical Press.)

is high; no special glassware is required, and no density gradient is formed. Cost is less than that of DGE, possibly as little as $1000, including power supply.

4. Sol—Gel Electrophoresis (SGE)

Gilman (40) performed model—cell separations (erythrocyte mixtures) on a conventional slab of agarose, which was melted during electrophoresis but gelled before and after. Cell electrophoresis experiments have been performed in the same all—glass vertical column shown in Figure 4 with and without a density gradient (41,42).

Model particles used in SGE experiments to date have consisted of rat, chicken, human, and rabbit erythrocytes fixed in 2.5% glutaraldehyde and 0.067 M sodium phosphate buffer, pH 7.2. Polymer materials were low—melting agaroses, such as SeaPrep and Sea Plaque (FMC Bioproducts, Inc.*) or standard agarose. Proof—of—principle experiments were carried out in 0.2% standard agarose at 56°C, a temperature that is incompatible with living cells and with many macromolecules. In these experiments a clean separation of rat and rabbit erythrocytes was achieved, and the agarose was re—solidified, thereby fixing the separated cells in place, as seen in Figure 6. These tests resulted in stable band formation and migration in the vertical column without a density gradient. SeaPrep (TM) was chosen as the low—melting agarose for tests in column electrophoresis, due to its compatibility with living cells (43), its ability to remain liquid at 32°C at 1% (m/v) concentration, and its miscibility with Ficoll at the concentrations used (42). Test particles consisting of a mixture of rat and rabbit erythrocytes were subjected to upward electrophoresis in a standard Ficoll gradient (41).

In all cases, large distances (1 cm) between separands were possible, implying that individual bands were free of cells from other populations. Purity counts were not made, but the photographs in Figure 6 imply that separated bands were 100% pure. Since the method of harvesting is slicing of the gel, the NSU is limited by the width of a gel slice. This is typically 2 mm from a possible transfer distance of 10 cm, so NSU = 50. In the language of chromatography, this method provides "baseline" separation, as band distortion due to pumping out the fractions does not occur. Band width measurements give CV = 5%. This method has not been practiced sufficiently widely to have provided adequate viability data; however, the cultivation of living animal cells in agarose is widely practiced (43). Low—melting agarose is absolutely essential, as high temperatures (namely, 56°C) kill animal cells. Droplet sedimentation may be somewhat retarded by the high viscosity of the agarose sol, so capacity is slightly higher than that of DGE. In horizontal SGE the sedimentation of cells to the bottom of the slab leads to awkward EEO problems, so sedimentation limits sample capacity in both orientations to approximately 10^7

---

*Use of this and other tradenames does not constitute endorsement by the National Institute of Standards and Technology. These are used only for accurate specification of procedures.

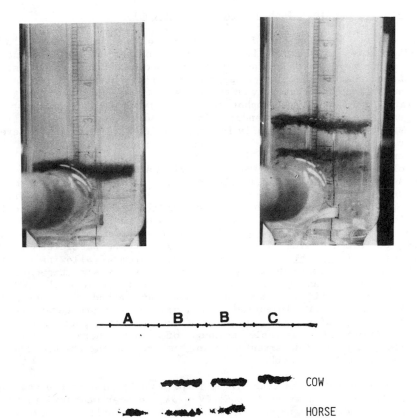

Figure 6. Agarose soil electrophoresis. Top panel: Photograph of separated bands of originally mixed fixed rat and rabbit erythrocytes at 0 (left) and after 1.0 h (right) of upward electrophoresis in 0.2% agarose at 56 °C (*41*). Bottom panel: Separation of horse (lane A) and cow (lane C) and originally mixed horse and cow (lanes b) erythrocytes in an agarose slab at 40 °C followed by resolidification (*40*). (Reproduced with permission from references 41 and 40. Copyrights 1988 and 1986 VCH Verlagsgeselischaft mbH.)

cells/hr.   In theory,   the slab SGE method should not be more
inconvenient   that   customary   slab gel electrophoresis.   In some
laboratories this could be a highly convenient method.   A medium-
quality apparatus for this method should cost about $2500.

5. Density Gradient Isoelectric Focusing (DGI)

Isoelectric focusing (IEF) is the movement of a separand through a pH
gradient to a pH at which it has zero net charge, at which point it
ceases to move through the separating medium.   The steady state
condition that occurs when a separand has reached equilibrium can be
characterized quantitatively on the basis of the dissociation constants
of water, the ampholytes (amphoteric electrolytes used to establish the
pH gradient) and the separands; this has been done is some detail by
Palusinski et al (44).   Equally interesting is the rate process whereby
water,   ampholytes   and   separands   approach   equilibrium   by
electrophoresis,   and   this   has   been   modeled   as   well   (45).   The
ampholytes   normally   used   to   establish   natural   pH   gradients   are
expensive and often harmful to living cells.   Non amphoteric, harmless
buffer systems have been developed to create artificial pH gradients
(46, 47).   This separation method is of limited use in the purification
of biological cells, most of which are isoelectric at a lower pH than
they can tolerate (48, 49, 50).   Nevertheless, experiments in DGI have
been   performed   using   living   cells   and   test   particles   (fixed
erythrocytes) (47, 51, 52, 53, 54).   The custom of allowing carrier
ampholytes and separands to approach steady state simultaneously in
ampholyte solutions at very low ionic strength for several hours is
quite lethal to living mammalian cells.   Thus methods were developed
for post–steady–state injection of cell samples at a pre–determined pH
point in the gradient (13, 49).   Resolution, capacity, purity, and cost
are   superior   to   comparable   methods   of   electrophoresis,   but   low
viability is a major deterrent to using this method for the preparative
separation of living cells.

Subpopulations of cells that differ in pI by 0.15 pH unit are resolved
(55).   NSU = 600, just as in DGE; CV = 5%, whether measured in terms
of migration distance or peak width in pH units.   Purity has been shown
to exceed 99% in model particle separations.   Viabilities tend to be
around 10%, which is unacceptable in a number of applications.   Owing
to four–fold increase in separation rate (see Figure 7) and a reduced
influence of droplet sedimentation, the capacity of DGI is greater than
that of DGE, namely around 3×10^7 cells/h.   The method is less
convenient than DGE, as both a density gradient and a pH gradient must
be established; however, there is a possibility that a system can be
contrived in which a density gradient is also an artificial pH gradient
(46, 56).   Capital cost is slightly higher than that of DGE, probably
around $2500.

6. Re–orienting Gradient Electrophoresis (RGE)

In column electrophoresis droplet sedimentation cannot be overcome
(13), and the only way to scale column electrophoresis is to increase
column cross–sectional area (14, 29).   Under conditions of droplet
sedimentation particles still move independently, and not collectively,

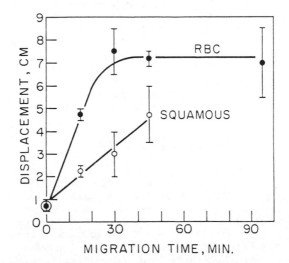

Figure 7. Migration plots (distance migrated vs. time) for human erythrocytes and exfoliated squamous epithelial cells from a uterine cervical specimen dispersed by gentle pipetting. The migrating squamous cell bands become broadened as a result of biological heterogeneity.   (Reproduced with permission from reference 48. Copyright 1977 Elsevier/North-Holland Biomedical Press.)

in the electric field. However, collective sedimentation during electrophoretic separation causes very broad bands at high feed concentrations. Rejecting heat from the top and bottom of a pancake-shaped column allows increased capacity in direct proportion to cross sectional area. This approach significantly reduces migration distance and hence NSU, but by re-orienting the cylinder after separation, the physical distance between bands of separands can be increased 4-5-fold, and the separands can be resolved by draining the column. A sketch of a RGE system is shown in Figure 8.

Like DGE, the RGE column is continuously sampled with a minimum fraction volume of 0.05 mL. Capacities up to 1 liter are achievable, so, conceivably NSU = 20000; in practice, as in other methods, CV = 6%, often more.   Purity values exceed 90% (further studies on model particle systems are needed), and viability = 90% is fairly routine. Capacity scales with area from DGE, so a 22 cm diameter RGE could process 5 x 10^8 cells/h.   The method is less convenient than conventional DGE, as re-orientation steps are added to the density gradient forming process. Availability is less easy to achieve, as a mechanical instrument must be built or purchased to perform slow, minimally disturbed re-orientation. Minimum cost of acquiring this technology, including chamber, electrodes and re-orienting device is around $5000.

7. Low-Gravity Column Electrophoresis (LGC)

The earliest electrophoresis experiments in low gravity were performed in a static cylindrical column on Apollo 14 and Apollo 16 lunar missions (57, 58, 59).   Extensive electroosmotic transport was observed, so that resolution was compromised. If the electroosmotic mobility μ_{eo} (coefficient of E in equation (10)) is high (of the order of 10^{-4} cm^2/V-s), then sample zones will be distorted into parabolas as shown in the simulation of Figure 2.   Also, an additive parabolic pattern is superimposed if there is a radial temperature gradient that causes a viscosity gradient. High temperature in the center of the chamber results in low viscosity and higher velocity due to both EEO return flow and faster migration of separands. Increased temperature also increases conductivity.   Excessive power input under typical operating conditions can result in 25% higher separand velocity in the center of the tube even in the absence of EEO (19). This undesirable, strictly thermal effect is independent of gravity, and it can be expected in low-gravity operations.

Adequate care was not taken to suppress EEO and to minimize radial thermal gradients in the Apollo electrophoresis experiments, and, although all gravitational effects were proved to be absent, EEO was comparable to the velocity of the test-particle separands (58, 59). Photographs taken during in-flight experiments were consistent with extreme examples of the parabolic distortions of the types modeled by Micale et al. (60, 61, 62) and shown in Figure 2.

A semi-automated static column electrophoresis system was designed for flight on the Apollo-Soyuz mission (63) and later used on space shuttle flight STS-3 (64, 65, 66). A sketch of its general features is given

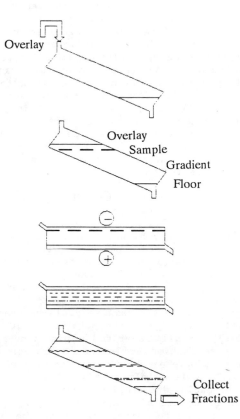

Figure 8.   Reorienting density gradient electrophoresis (14,29).
The flat cylindrical chamber (1.5–3.0 cm high x 10–20 cm
diameter) is tilted for filling at the bottom corner, with
sample layered at the top of a linear density gradient. The
field is applied with the chamber in the horizontal position .
After electrophoretic separation the chamber is reoriented for
the collection of fractions. Courtesy of R. M. Stewart.

in Figure 9. The system design included a procedure for minimizing EEO by silylating the heavily cleaned glass chamber walls and coating them with methylcellulose (67). A low-conductivity buffer "A-1" (63) was developed to minimize current density per unit field and hence minimize Joule heating. No protein separations were done with this system, but several cell types were separated on the Apollo-Soyuz flight. Sample cells were frozen in a delrin disk that was inserted into the electrophoresis columns. Columns were 6 mm in diameter and 15 cm long, to accommodate particles with $\mu = -1.0$ to -3.0×10^{-4} cm^2/V-s during a 1-h separation. Temperature control was provided by a thermoelectric cooler operated in two modes: thermostating at 25°C during electrophoresis and freezing the entire column after completion of each run. Each column could then be stored frozen for return to earth with separated cells at different positions in the column to be sliced (like a loaf of bread) while still frozen for the collection of separated cell subpopulations. All of the electrophoresis fluids contained glycerol for the cryopreservation of living and fixed cells. Aldehyde-fixed erythrocytes (68) were used as test particles. Despite severely attenuated viability after flight, it was possible to obtain electrophoretic mobility distributions and unique subpopulations of cultured human kidney cells (69, 28).

The same apparatus was used in Space Shuttle flight STS-3 in an experiment designated "EEVT" and designed to confirm the kidney cell separations achieved on Apollo-Soyuz with 6 more separations and to test the hypothesis that very high concentrations of test particles (fixed erythrocytes) can be separated in low gravity owing to the lack of particle and zone sedimentation. Gravity-independent effects such as cell-cell interaction, reduced conductivity, and electrostatic repulsion could be sought in low-gravity experiments at high cell concentration (70). Thanks to in-flight photography during separations, the two erythrocyte samples could be tracked through the course of a 1-h separation (17) as shown in Figure 10. Test particles (fixed erythrocytes) generally separated at high concentration on orbit separated in the same fashion as low-concentration cells in normal gravity. However, a single parabolic pattern of human and rabbit red cells is seen migrating from left to right, when two distinct bands were expected. The leading and trailing edges corresponded to the μ's of human and rabbit red cells, respectively (17, 66). Simulation experiments in a rotating tube indicated that the observed pattern could occur if the human:rabbit cell ratio was 2:1 and a significant (but undetermined) EEO was present (16). Each frozen tube was sliced into 20 sections, so NSU = 20. In the experiments with fixed erythrocytes on the ASTP mission the CV was 10%. Bands overlapped, and the maximum recorded purity was 87%. The viability of the cultured kidney cells recovered from the frozen tubes was <10%. The capacity was limited only by the volume of cells that could be applied, and on STS-3 this was as high as 10^9 cells, and processing time was 1 hr, so capacity = 10^9 cells/h. There is no need to elaborate on the convenience of conducting experiments in orbital space flight. The cost per 10 separations was about $300000.

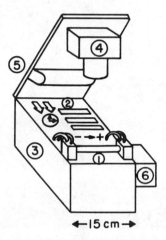

←|5 cm→

Figure 9. Sketch of free zone electrophoresis apparatus used on Apollo–Soyuz Test Project and Space Shuttle flight STS–3. The hinged cover held an illuminator (5) and a camera (4) that viewed the clock, voltmeter, ammeter (2) and, when uncovered, the electrophoresis column (1). Columns were interchangeable and were served by connectors to electrode compartments connected to a buffer recirculation system inside the housing (3). An accelerometer (6) was attached for the Space Shuttle mission.

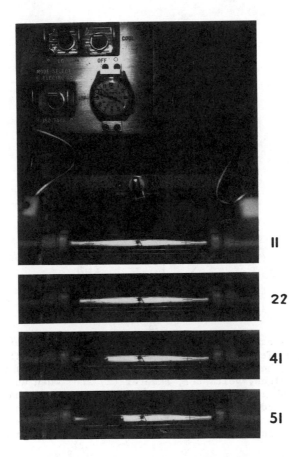

Figure 10. Photographs showing top panel of free zone
electrophoresis unit during Space Shuttle mission during 51
minutes of separation of a test mixture of concentrated human
and rabbit erythrocytes. The column cooling cover was removed,
and these photographs were made at 11, 22, 41 and 51 minutes
(top to bottom, respectively). A broad band of cells is visible
moving from left to right (17,20,66). (Reproduced with permission
from reference 20. Copyright 1990 American Institute of Aeronautics and
Astronautics.)

8. Continuous Free Flow Electrophoresis (FFE)

Free flow electrophoresis, perhaps the most popular electrokinetic method of cell separation, has not found favor as an industrial–scale purification process. The process cannot be scaled up because convection occurs as a consequence of inadequate heat removal from current–carrying electrolyte solutions, and sample zone convection occurs when sample concentration is too high. Various approaches to this problem have been utilized. Conventional commercial free–flow electrophoretic separators have used downward flow and brine cooling to counteract the two above problems. Or, in at least one case, upward flow over a long distance (>100 cm) with low field strength (<20 V/cm) and use of the electrode buffer as a coolant counteract these two problems. A generic free flow separator is sketched in Figure 11.

A continuous flow electrophoresis system is sketched diagrammatically in Figure 9. Several investigators have constructed mathematical models of FFE (71, 72, 73). The main ingredients of such a model include the following variables (74): electrophoretic migration of separands, Poiseuille flow of pumped carrier buffer, electroosmosis at the front and back chamber walls, horizontal and vertical thermal gradients, conductivity gradients, diffusion, sample input configuration, and fraction collection system. Figure 12 indicates the path followed by three categories of separands, low, medium and high mobility and indicates the effects of Poiseuille flow and electroosmosis when sample is injected at the bottom, electrophoresis is to the right, and fractions are collected at the top. A coordinate system is shown in which the z axis is the direction of pumped fluid (buffer) flow, the y axis is the chamber thickness, and the x axis is the chamber width and direction of migration of separands in the applied, horizontal electric field. Typical dimensions of available FFE chambers is x_m = 6–15 cm, y_m – 0.03–0.15 cm (this dimension can be increased to 0.30 cm in low gravity), and z_m – 22–110 cm, where the subscript m designates distance to the opposite wall from the origin. The motion of a particle through such a chamber is complex. The migration distance x is the sum of electrophoretic migration and electroosmotic flow

$$x = \mu E \tau (y) + V_{eo}(y)\, \tau (y) \qquad (14)$$

where τ = residence time of the separand in the column. But $\tau(y)$ and $v_{eo}(y)$ are explicit functions of y (75). Under some conditions (non-uniform conductance) E is also a function of x and y (76). Separands arrive at the outlet end of the chamber distributed in nested crescents in which separand particles nearest the front and back walls migrate the farthest, as indicated in Figure 13. However, the y-dependencies can be minimized by using a thick chamber and limiting the sample stream to a central zone near the midpoint of the chamber in the y direction. A thicker chamber is allowed in low gravity where thermal distortions of flow presumably do not occur, ignoring Marangoni flow due to a small thermal gradient in the z direction and the viscosity gradient mentioned above. Figure 13 indicates how sharper bands can be achieved in a thicker chamber. However, such bands can be unsharpened by electrohydrodynamic phenomena when the conductivity of

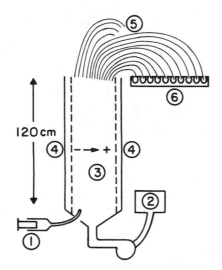

Figure 11.   Sketch of generalized continuous flow
electrophoresis (FFE) system. Sample stream (1) enters chamber
(3) near the bottom (in this case) of the upward-flowing, pumped
(2) buffer system. The component separands migrate from left to
right in the electric field between electrodes (4) according to
their electrophoretic mobility and exit at the top of the
chamber through individual outlets, 20-200 in number (5),
depending on chamber design, for collection into fraction tubes
(6). (Reproduced with permission from reference 16. Copyright 1985
Walter de Gruyter.)

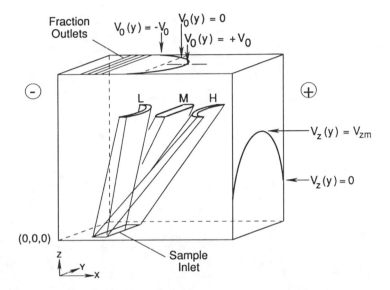

Figure 12.   Schematic diagram (not to scale) of an upward-flowing free-flow electrophoresis chamber. A coordinate system with (0,0,0) at the lower front left corner is shown. Sample stream is shown as a slit (circular sample streams are commonly used). Anodal migration of low, medium, and high (L,M,H) mobility separands results in crescent-shaped bands due to Poiseuille retardation of particles near the walls (where $v_z(y)$ is low) in the case of high-mobility and electroosmotic flow near the walls (where $v_o(y)$ is high) in the case of low mobility. These two parabolic distortions balance in the case of medium-mobility separands. (Reproduced with permission from reference 20. Copyright 1990 American Institute of Aeronautics and Astronautics.)

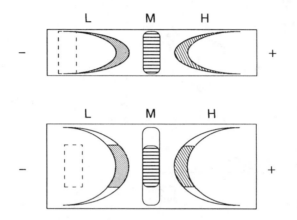

Figure 13. A thicker FFE chamber allows the separation of separands in narrower bands, resulting in higher output purity. Preventing the sample stream from approaching the front and back chamber walls results in less band distortion due to Poiseuille and electroosmotic flow. Dashed rectangle represents a cross section of the original sample stream at the inlet. (Reproduced with permission from reference 20. Copyright 1990 American Institute of Aeronautics and Astronautics.)

the sample zone is not properly matched to that of the carrier buffer, resulting in E(x,y) not being constant, because k(x,y) is not constant (64, 76, 77, 78). This phenomenon is also independent of gravity, although any effect of g on heat transfer will also affect k(x,y), as k depends linearly on temperature. The general relationship for field distortion is derived from Taylor's "electrohydrodynamic" concept applied to a spherical drop (79); it assumes a sample zone of circular cross section small compared to the y dimension of the chamber and having different conductivity k_s, viscosity η_s, and dielectric constant ϵ_s from that of the carrier buffer. The surface stress S is given by

$$S = \frac{E^2 e \left[e/e_s \{ (k_s/k)^2 + (k_s/k) + 1 \} - 3 \right]}{4\pi \{ (k_s/k)^2 + 1 \}^2} cos2\Theta \qquad (15)$$

where Θ is the angle subtended by the stress vector and the applied electric field lines. This function predicts that when $k_s > k$ the sample stream will be pulled outward at $\Theta = 90°$ (toward the chamber walls) and compressed in the perpendicular direction, as in the upper example of Figure 14, and vice versa as in the lower example.

Approximate relationships can be derived for the overall motion of a separand particle in FFE by substituting explicit geometrical functions into equation (14) to give

$$x(y) = \frac{2Ez_m(\mu - \mu_{eo}/2)}{3V_{zav}[1 - (y-b^2/b^2]} \qquad (16)$$

which states that the direction of the parabola depends on the relative magnitudes of the separand mobility and the electroosmotic mobility, μ_{eo}, of the chamber wall (80), where the geometrical variables are defined in Figure 12. In the example illustrated in Figure 12, high-mobility separands ("H") migrate farthest to the right near the chamber walls, while low ("L") or zero-mobility separands move farthest to the left near the chamber walls. The effects of temperature gradients on conductivity and viscosity can be added (80, 81).

The above relationships deal only with the gravity-independent components of separand motion. Gravity-dependent variables are treated more qualitatively, and the critical Rayleigh number,

$$Ra_c = \frac{g\beta |\Delta T| x_m^3}{\alpha \nu} < \pi^4 \qquad (17)$$

must generally not be exceeded to prevent convective instability due either to solutal zone sedimentation (equation (3)) or thermal convection (equation (23)), where β is thermal expansion coefficient, α is thermal diffusivity, and ν is kinematic viscosity (4, 72, 82). Dimensionless analysis by Ostrach (82) and Saville (72) predicted asymmetric stationary instabilities (and corresponding temperature profiles) at lower Ra when carrier buffer and coolant are both flowing downward so that the chamber is warmer at the bottom than at the top. Deiber and Saville (83) pointed out that the cathodic side of the chamber will be more effectively cooled by the action of electroosmosis, which sweeps carrier buffer inward from the walls at the cathode side and outward from the center at the anode side. This coupling of convection to electroosmosis results in predicted asymmetric flow distortions, which have been observed (84). Thermal distortions of flow depend upon gravity and are sensitive to chamber thickness (x_m^3). At 1 g practical chamber thicknesses are limited to about 0.5 mm. Ivory (4) has calculated that by allowing the chamber to heat to 40°C when g = 0, greater chamber thickness is feasible, and a 440—fold increase in sample capacity could be realized in FFE. This is similar to the experimentally demonstrated enhancements in protein separations in low gravity (11). Some stability can be gained by concurrent flow of carrier buffer and coolant upward, so that a small temperature rise (ca. 3°C over 100 cm column length) reduces the density at the top of the column (11).

FFE units have been constructed with 20 to 197 outlets, so at the current state of the art, NSU = 197. The corresponding theoretical CV is 0.5%, which has never been achieved. From the data in Figure 15 a measured CV for a mixture of fixed erythrocytes could be determined, giving CV = 7.1% (28). From the same data, maximum purity was found to be 98.6% for the more rapid component and 98.9% for the lower—mobility component of the mixture. High viability for mouse lymphocytes, around 96%, has usually been claimed for this method when low—conductivity triethanolamine buffers are used (85). As a continuous electrophoresis system, FFE should have higher capacity than the batch methods discussed above. The product of maximum cell concentration and maximum feed flow rate is 10^7 cells/mL x 10 mL/h = 10^8 cells/h. It is difficult to realize this rate owing to sample zone sedimentation, which prevents the sample from entering upward buffer flow and reduces electrophoretic migration distance in downward flow. Typical separations have about 1/10th this capacity. Some aspects of FFE make it convenient: no density gradient, continuous sample collection, potential for real—time optical monitoring. Access to apparatus is realistic, and commercial versions of FFE can be purchased at prices ranging from $20000 to $60000.

9. Continuous Flow Isoelectric Focusing (CFI).

The inlet buffers in FFE can be contrived to form various gradients in conductivity, pH etc. Just et al. (86, 87) introduced mixed ampholytes into a downward—flowing FFE and injected a stream of living erythrocytes so that the ampholytes and the cells focused simultaneously during a single pass through the FFE. It was found difficult to create conditions consistent with cell viability, as erythrocytes are isoelectric around pH 2.2 (88).

The apparatus used for the published experiments had 48 outlets, so NSU = 48. The focusing bandwidth was 4 channels out of a total of 28 migrated, giving CV = 7.4%. With appropriate pH gradients, it is assumed that 100% purity is achievable. There appear to be no viabilities reported for CFI experiments—viability may be no better than in DGI, around 10%. The capacity is subject to the same constraints as in FFE, so 10^8 cells/h must be considered an upper limit. CFI is less convenient than FFE, as ampholyte focussing must be considered in addition to the usual features of FFE. The same equipment is used, with the same availability and cost.

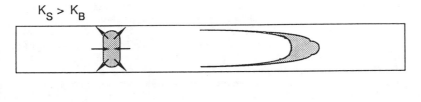

Figure 14.   Electrohydrodynamic distortion of sample bands in
FFE.   From equation (21) the stress vector at the "surface" of a
high-conductivity sample stream is outward perpendicular to the
electric field and inward parallel to the field lines (top
diagram) and vice versa for a sample stream with lower
conductivity than that of the carrier buffer (lower diagram).
(Reproduced with permission from reference 20. Copyright 1990 American
Institute of Aeronautics and Astronautics.)

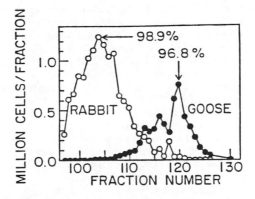

Figure 15.   Test-particle separation by CFE.   Electrophoretic
migration pattern of a mixture of fixed rabbit and goose red
blood cells in triethanolamine buffer.   Migration was from left
to right. (Reproduced with permission from reference 28. Copyright 1989
CRC Press, Inc.)

10. Rotating Annular Electrophoresis (RAE)

In one system the buffer curtain flows upward in the annular zone
between a vertical stationary and a rotating cylinder. This system and
conventional FFE systems have been compared by Mosher et al. (5). No
living cell separations with this system have been reported. The
device as designed has 19 outlets, so NSU = 19, and CV can never be
less than 5.5%. Shear stresses exceeding $10^{-5}N/cm^2$ are encountered,
and these could be lethal to living nucleated mammalian cells (89).
The apparatus is designed for maximum capacity, and several liters per
hour can be processed. Convenience is somewhat compromised by the
amount of feed and buffer that must be prepared. Access is extremely
difficult, as the sole source has discontinued manufacturing the
apparatus, which is difficult to build. When it was available the
retail price was greater than $100,000.

11. Low Gravity Continuous Flow Electrophoresis (LGF)

One heroic method of FFE uses the low-gravity environment of space
flight (11, 90). Of these two high-capacity methods, RAE sacrifices
resolution, and LGF is costly and seldom available. Free-flow
electrophoresis in low gravity, along with electrophoresis in low
gravity in general, was probably born as a concept in 1969 (74). This
separation method was first implemented in spaceflight in an experiment
named "MA-014," 16 July, 1975, on the Apollo-Soyuz Test Project (ASTP).
The stated purpose of this experiment was "to verify the theoretically
expected better and higher capacity separations in the free-flow
electrophoresis system under zero-g conditions" (90). Specific
objectives included the characterization of temperature and velocity
effects in a wide gap, the separation of preparative quantities of
living cells, and an engineering study for potential future Spacelab
equipment. To fit the envelope of the crowded spacecraft, the fully
automated system served a separation chamber that was 3.8 mm thick (not
considered feasible at 1 x g) by 28 mm wide by 180 mm long (electrode
length). No fractions were collected, but the system was monitored
optically using a halogen lamp and a scanning detection system to
measure light absorbance with a 128-sector photodiode array. All test
samples were cell suspensions (rat bone marrow, mixed human and rabbit
erythrocytes (RBC), rat spleen, and rat lymph node with human RBC
markers. Starting samples approaching 10^8 cells/mL were used in all
cases, and separations were compared with 1-g counterparts at lower
cell concentrations (0.1 times those used in spaceflight (91)). With
one exception (bone marrow), distributions were essentially the same
in low gravity and at 1 g. Effects of low gravity on resolution and
throughput were not quantitated.

McCreight et al. (74) designed and built a low-gravity free-flow
electrophoresis unit for sounding-rocket flight. It was to perform
electrophoretic separations during the 7-min low-gravity mission of a
Black Brant sounding rocket (1973). Although only 5-7 min of low
gravity is a short time, a small chamber, such as that used by Hannig

et al., was considered capable of functioning and performing separations. Chamber dimensions were 5 mm thick by 5 cm wide by 10 cm long (electrode length). The design includes an optical scanner and an arrangement for collecting up to 50 fractions, giving a collection resolution of 2% and hence an ability to measure widths of bands of separands.

Snyder et al. (64, 76) were able to demonstrate, qualitatively, behavior of particulate separands in sample streams with conductivity unmatched to carrier buffer under low-gravity conditions, where convective disturbances played no role, and under laboratory conditions (AC field) where separation did not occur but photography was feasible. They concluded: "Additional effort is also needed to determine the role of particle concentration, if any, in free-flow electrophoretic separations. Laboratory experiments clearly show limitations that are not entirely due to droplet sedimentation." In any case discontinuities in buffer ion concentrations have been shown to produce electrohydrodynamic distortions of sample bands that seriously compromise separations, as illustrated in Figure 14. Soluble separands that add significantly to the conductivity of sample streams at high concentrations would presumably have the same effect. It remains to determine whether or not particulate separands would, at high concentration, also cause such field distortions. Because very highly concentrated fixed erythrocytes migrated at expected velocities in the STS-3 experiment cited above, there exists one piece of evidence that particulate separands will not induce, by themselves, electrohydrodynamic distortion during electrophoretic separation.

12. Stable Flow Density Gradient Electrophoresis (STF)

The anticonvective stability provided by a density gradient (as in DGE) is combined with the capacity of continuous flow electrophoresis (FFE) in the "Stable Flow Free Boundary Electrophoresis" (STAFLO) system (92). Buffers having 12 different densities are layered at the entrance of an horizontal rectangular pipe approximately 5 by 5 cm by 25 cm long. A vertical electric field is applied between electrode compartments above and below the flow chamber and separated from the flow chamber by membranes. The sample is injected into one of the 12 flow channels, and separands are deflected upward or downward, according to their μ_e, and they arrive in one of the 12 channels at the downstream end of the chamber.

Because there are 12 channels NSU = 12, and the minimum CV = 8.5%. Overlap of separands between adjacent channels in this system is inevitable (93), so purity >90% is rare. The system tolerates a wide range of buffers, so viability = 100% is possible (94). Although the processing rate may be as high as 25 mL of sample per 10 min, the thin layers are susceptible to droplet sedimentation (95), and the capacity is estimated at 9×10^7 cells/h. Forming the flowing gradient requires the preparation of 12 buffers with different densities, so use of the system is moderately inconvenient. Access to existing systems is possible through only 3–4 laboratories in the west, where it is used as a hematology research tool (94, 96), but interest in this technology is rising in the U.S.S.R. A custom-built unit can be made by an

Table I. Comparison of methods of cell electrophoresis using absolute values

NO.	METHOD	RESOLUTION NSU	CV%	VIABIL. %	PURITY %	CAPACITY CELLS/H	CONVENIENCE USE	ACCESS	CAPITAL COST,K$
1	FZE	30	5	100	99.8	20×10^7	MOD	LOW	30–50
2	DGE	600	5	85	99.6	0.5 "	MOD	HIGH	2.0–2.5
3	AVE	120	?	90	>91	0.7 "	HIGH	HIGH	1.5–2.0
4	SGE	50	4–5	?	100	1.0 "	HIGH	HIGH	2.0–2.5
5	DGI	600	5	<20	99+	3.0 "	MOD–	MOD	2.5–3.0
6	RGE	>10000	>6	90	90	50 "	MED	LOW	4.5–5.5
7	LGC	20	10	<10	87	100 "	LOW	ZERO	>300
8	FFE	197	7	>90	98.9	10 "	MOD	LOW	20–60
9	CFI	48	7+	10	99+	10 "	LOW	LOW	20–60
10	RAE	19	>5.5	?	?	10000 "	MOD	V.LOW	>100
11	LGF	197	7	55	99	>10	V.LOW	ZERO	>10000
12	STF	12	>8	>90	<80	5	MOD	MOD	6–20
13	CBE	20	5	85	99	50	MOD	LOW	50–100

Table II. Comparison of methods of cell electrophoresis using figures of merit

NO.	METH.	RESOL.	VIABIL.	PURITY	CAPACITY	CONVENIENCE	COST	TOTAL
1	FZE	3+	4	4	3	1	2	17+
2	DGE	4	4	4	2	3	4	21
3	AVE	1	3	3	3	4	4	18
4	SGE	3	3	3	3	4	4	20
5	DGI	4	1	4	2	2	4	17
6	RGE	3	4	3	4	3	3	20
7	LGC	4	2	3	4	0	0	13
8	FFE	3	4	2	3	2	2	16
9	CFI	?	1	3	3	2	2	11+
10	RAE	1	1	?	4	2	1	9+
11	LGF	3+	2	4	4	0	0	13+
12	STF	2	4	2	3	2	4	17
13	CBE	3	4	3	3	2	2	17

institutional machine shop for $5000–12000 including the price of a power supply, so the capital cost and convenience are comparable to those of RGE.

13. Endless Continuous Belt Electrophoresis (EBE)

In a process called "Endless Belt Continuous Flow Deviation Electrophoresis" Kolin (97, 98) designed a horizontal annular vessel surrounding a bar magnet. The horizontal passage of a current through the buffer in the annulus perpendicular to the radial field lines of the magnet creates a steady flow of the buffer around the annulus due to the Lorentz force. This flow mimics the rotation of the horizontal tube in FZE (see above) thereby counteracting both sedimentation and the build–up of convective vortices. The annular space is typically 1.5 mm thick. The sample is introduced continuously at the appropriate end of the annulus, and separands migrate in helical paths the pitch of which is directly proportional to μ_e. After an adequate number of helical turns (about 5) are accomplished, separands are removed from the annulus at the opposite end through a linear array of individual ports, typically 20, so NSU = 20, and minimum CV = 5%. A field strength up to 150 V/cm is possible, so that migration is rapid. Zone sedimentation is avoided by the buffer circulation, and overall capacity is claimed to be 5×10^8 cells/h (98). Buffers of around 0.6 mS/cm conductivity are used, and viabilities around 85% can be expected. Photographs indicate practically no overlap between bands of separands during migration, so purity should be around 99%.

Summary:

Table I lists the 13 methods of cell electrophoresis considered in this study and the absolute values of the 8 parameters chosen for evaluation. Table II lists the corresponding figures of merit for the different methods. These figures were derived from an approximately linear interpretation of the data in Table I. Overall, a simple, accessible, inexpensive method, if it can provide viable cell populations is the reversible gel method, since related techniques are practiced in many laboratories daily. On the other hand, free flow electrophoresis is the most widely practiced continuous method. Its capacity and advanced stage of development, including commercialization, rank it high among processes that are useful in laboratories that wish to separate cells daily.

Acknowledgments:

This work was supported by the National Aeronautics and Space Administration (Contracts NAS9–15583, NAGW–694, and NAS9–17431), U. S. Public Health Service (Grants R01 CA–12589 and N01 CB–43984) and the National Institute of Standards and Technology. Experimental and theoretical work by Lindsay D. Plank, M. Elaine Kunze, Kenneth D. Cole and Robin Stewart and the guidance of F. J. Micale and Scott R. Rudge are gratefully acknowledged. Senior design students of the Colorado School of Mines are acknowledged for their contribution to the study of re–orienting density gradient electrophoresis.

Literature Cited:

1. Hunter, R. J. Zeta Potential in Colloid Science, Principles and Applications; Ottewill, R. H.; Rowell, R. L. Eds.; Academic Press: New York, NY, 1981.
2. O'Brien, R. W.; White, L. R. J. Chem. Soc. Faraday II 1978, 74, 1607-1626.
3. Jorgensen, J. W.; Lukacs, K. D. Anal. Chem. 1983, 53, 1298.
4. Ivory, C. F. Sep. Sci. and Technol. 1988, 23, 875-912.
5. Mosher, R.; Thormann, W.; Egen, N. B.; Couasnon, P.; Sammons, D. W. In New Directions in Electrophoretic Methods; Jorgenson, J. W.; Phillips, M. Eds.; American Chemical Society: Washington, DC, 1987; pp 247-262.
6. Zeiller, K.; Hannig, K. Hoppe-Zeylers Z. Physiol. Chem. 1971, 352, 1162-1167.
7. McLaughlin, S; Poo, M.M. Biophys. J. 1981, 34, 85-93.
8. Zimmerman, U. Rev. Physiol. Biochem. Pharmacol. 1986, 105, 175-252.
9. Todd, P.; Plank, L. D.; Kunze, M. E.; Lewis, M. L.; Morrison, D. R.; Barlow, G. H.; Lanham, J. W.; Cleveland, C. J. Chromatography 1986, 364, 11-24.
10. Plank, L. D.; Hymer, W. C.; Kunze, M. E.; Todd, P. J. Biochem. Biophys. Meth. 1983, 8, 273-289.
11. Hymer, W. C.; Barlow, G. H.; Cleveland, C.; Farrington, M.; Grindeland, R.; Hatfield, J. M.; Lanham, J. W.; Lewis, M. L.; Morrison, D. R.; Rhodes, P. H.; Richman, D.; Rose, J.; Snyder, R. S.; Todd, P.; Wilfinger, W. Cell Biophysics 1987, 10, 61-85.
12. Mason, D. W. Biophys. J. 1976, 16, 407-416.
13. Boltz, R. C. Jr.; Todd, P. In Electrokinetic separation Methods; Righetti, P. G.; van Oss, C. J.; Vanderhoff, J. W., Eds.; Elsevier/North-Holland Biomedical Press: Amsterdam, 1979; pp 229-250.
14. Tulp, A.; Timmerman, A.; Barnhoorn, M. G. In Electrophoresis '82; Stathakos, D., Ed.; W. deGruyter & Co.: Berlin, 1983; pp 317-323.
15. Todd, P.; Hjertén, S. In Cell Electrophoresis; Schütt, W; Klinkmann, H., Eds.; Walter deGruyter & Co.: Berlin, 1985; pp 23-31.
16. Todd, P. In Cell Electrophoresis; Schütt, W.; Klinkmann, H., Eds.; W. deGruyter & CO.: Berlin, 1985; pp 3-19.
17. Snyder, R. S.; Rhodes, P. H.; Herren, B. J.; Miller, T. Y.; Seaman, G. V. F.; Todd, P.; Kunze, M. E.; Sarnoff, B. E. Electrophoresis 1985, 6, 3-9.
18. Omenyi, S. N.; Snyder, R. S.; Absolom, D. T.; Neumann, A. W.; van Oss, C. J. J. Colloid Interface Sci. 1981, 81, 402-409.
19. Hjertén, S. Free Zone Electrophoresis; Almqvist and Wiksells Boktr. AB: Uppsala, 1962.
20. Todd, P. In Progress in Low Gravity Fluid Dynamics and Transport Phenomena; Koster, J. N.; Sani, R. L., Eds. American Institute of Aeronautics and Astronautics: Washington, 1990; in press.
21. Abramson, H. W.; Moyer, L. S.; Gorin, M. H. Electrophoresis of Proteins; Reinhold Publ. Corp: New York, NY, 1942.

22. Bangham, A. D.; Flemans, R.; Heard, D. H.; Seaman, G. V. F. Nature 1958, 182, 642.
23. Brinton, C. C., Jr.; Laufer, M. A. In Electrophoresis; Bier, M., Ed.; Academic Press: New York, 1959; pp 427–492.
24. Seaman, G. V. F. In Cell Electrophoresis; Ambrose, E. J., Ed.; Little, Brown & Co.: Boston, MA, 1965, pp 4–21.
25. Hjertén, S. In Cell Separation Methods; Bloemendal, H., Ed.; Elsevier/North–Holland Biomedical Press: Amsterdam, 1977; pp 117–128.
26. Poulson, A.; Cramer, R. Biochim. Biophys. Acta 1958, 29, 187–192.
27. Boltz, R. C. Jr.; Todd, P.; Gaines, R. A.; Milito, R. P.; Docherty, J. J.; Thompson, C. J.; Notter, M. F. D.; Richardson, L. S.; Mortel, R. J. Histochem. Cytochem. 1976, 24, 16–23.
28. Todd, P.; Kurdyla, J.; Sarnoff, B. E.; Elsasser, W. in Frontiers in Bioprocessing; Sikdar, S. K.; Bier, M.; Todd, P., Eds.; CRC Press: Boca Raton, Fl, 1989; 223–234.
29. Tulp, A. Methods Biochem. Anal. 1984 30, 141–198.
30. Griffith, A. L.; Catsimpoolas, N.; Wortis, H. H. Life Sci. 1975, 16, 1693–1702.
31. Platsoucas, C.D.; Good, R.A.; Gupta, S. Proc. Natl. Acad. Sci. U.S.A. 1979, 76, 1972–1976.
32. Platsoucas, C.D.; Beck, J.D.; Kapoor, N.; Good, R.A.; Gupta, S. Cell. Immunol. 1981, 59, 345–354.
33. Boltz, R. C. Jr.; Todd, P.; Streibel, M. J.; Louie, M.K. Preparative Biochem. 1973, 3, 383–401.
34. Brooks, D. E.; Seaman, G. V. F. J. Coll. Interface Sci. 1973, 43, 670–686.
35. Todd, P.; Hymer, W. C.; Plank, L. D.; Marks, G. M.; Hershey, M.; Giranda, V.; Kunze, M. E.; Mehrishi, J. N. in Electrophoresis '81; Allen, R. C.; Arnaud, P., Eds.; W. de Gruyter & Co.: New York, NY, 1981; pp 871–882.
36. Plank, L. D.; Kunze, M. E.; Todd, P. Submitted (1989). Gaines, R. A. Thesis. The Pennsylvania State University: University Park, Pennsylvania, 1981.
37. Plank, L. D.; Todd, P.; Kunze, M. E.; Gaines, R. A. Electrophoresis '81, Book of Abstracts, 1981, 125.
38. Gillman, C. F.; Bigazzi, P. E.; Bronson, P. M.; Van Oss, C. J. Prep. Biochem. 1974, 4, 457–472.
39. van Oss, C. J.; Bronson, P. M. In Electrokinetic Separation Methods; Righetti, P. G.; van Oss, C. J.; Vanderhoff, J. W., Eds.; Elsevier/North–Holland: Amsterdam, 1979; pp 251–256.
40. Gilman, R. Electrophoresis 1986, 7, 41–43.
41. Plank, L. D.; Kunze, M. E.; Gaines, R. A.; Todd, P. Electrophoresis 1988, 9, 647–649.
42. Todd, P.; Szlag, D. C.; Plank, L. D.; Delcourt, S. G.; Kunze, M. E.; Kirkpatrick, F. H.; and Pike, R. G. Adv. Space Res. 1989, 9 (11), 97–103.
43. Nilsson, K.; Scheirer, W.; Katinger, H. W. D.; Mosbach, K. Methods of Enzymol. 1987, 135, 399–410.
44. Palusinski, O. A.; Allgyer, T. T.; Mosher, R. A.; Bier, M.; Saville, D. A. Biophys. Chem. 1981, 14, 389–397.
45. Palusinski, O. A.; Bier, M.; Saville, D. A. Biophys. Chem. 1981, 14, 389–397.

46. Troitsky, G. V.; Azhitsky, G. Yu. Isoelectric focussing of proteins in natural and artificial pH gradients. Kiev Nauka Dumka: Kiev, 1984.
47. Boltz, R. C., Jr.; Miller, T. Y.; Todd, P.; Kukulinsky, N. E. In Electrophoresis '78; Catsimpoolas, N., Ed.; Elsevier/North-Holland Biomedical Press: Amsterdam, 1978; pp 345-355.
48. Boltz, R. C., Jr.; Todd, P.; Hammerstedt, R. H.; Hymer, W. C.; Thompson, C. J.; Docherty, J. J. In Cell Separation Methods; Bloemendal, H., Ed.; Elsevier/North-Holland Biomedical Press: Amsterdam, 1977; pp 145-155.
49. Sherbet, G. V. The Biophysical Characterisation of the Cell Surface; Academic Press: London, 1978.
50. McGuire, J. K.; Miller, T. Y.; Tipps, R. W.; Snyder, R. S.; Righetti, P. G. J. Chromatog. 1980, 194, 323-333.
51. Leise, E.; LeSane, F. Prep. Biochem. 1974, 4, 395-410.
52. Sherbet, G. V.; Lakshmi, M. S.; Rao, K. V. Exp. Cell Res. 1972, 70, 113-123.
53. Hammerstedt, R. H.; Keith, A.D.; Boltz, R. C., Jr.; Todd, P. Arch. Biochem. Biophys. 1979, 194, 565-580.
54. Zarkower, D.; Plank, L. D.; Kunze, M. E.; Keith, A.; Todd, P; Hymer, W. C. Cell Biophys. 1984, 6, 53-66.
55. Thompson, C. J.; Docherty, J. J.; Boltz, R. C., Jr.; Gaines, R. A.; Todd, P. J. Gen. Virol. 1978, 39, 449-461.
56. Cole, K. D., Dutta, B. K. and Todd, P. Non-amphoteric isoelectric focusing III. A borate-polyol density gradient for rapid isoelectric focusing of proteins. In preparation, 1990.
57. Snyder, R. S. Electrophoresis demonstration on Apollo 16. National Aeronautics and Space Administration Report NASA TMX-64724; National Aeronautics and Space Admin., Huntsville, AL, 1972.
58. Snyder, R. S.; Bier, M.; Griffin, R. N.; Johnson, A. J.; Leidheiser, H.; Micale, F. J.; Ross, S.; van Oss, C. J. Sep. Purif. Meth. 1973, 2, 258-282.
59. McKannan, A. C.; Krupnick, E. C.; Griffin, R. N.; McCreight, L. R. National Aeronautics and Space Administration Report NASA TMX-64611; National Aeronautics and Space Admin., Washington, DC, 1971.
60. Micale, F. J.; Vanderhoff, J. W.; Snyder, R. S. Sep. Purif. Meth. 1976, 5, 361-383.
61. Vanderhoff, J. W.; Micale, F. J. In Electrokinetic Separation Methods; Righetti, P. G.; Van Oss C. J.; Vanderhoff, J. W., Eds.; Elsevier/North-Holland: Amsterdam, 1979; pp 81-93.
62. Vanderhoff, J. W.; van Oss, C. J. In Electrokinetic Separation Methods; Righetti, P. G.; van Oss, C. J.; Vanderhoff, J. W., Eds.; Elsevier/North-Holland: Amsterdam, 1979; pp 257-274.
63. Allen, R. E.; Rhodes, P. H.; Snyder, R. S.; Barlow, G. H.; Bier, M.; Bigazzi, P. E.; van Oss, C. J.; Knox, R. J.; Seaman, G. V. F.; Micale, F. J.; Vanderhoff, J. F. Sep. Purif. Meth. 1977, 6, 1-59.

64. Snyder, R. S.; Rhodes, P. H.; Miller, T. Y.; Micale, F. J.; Mann, R. V.; Seaman, G. V. F. Sep. Sci. Technol. 1986, 21, 157–185.

65. Morrison, D. R.; Lewis, M. L.; In 33rd International Astronautical Federation Congress; 1983, Paper No. 82–152.

66. Sarnoff, B. E.; Kunze, M. E.; Todd, P. Adv. Astronaut. Sci. 1983, 53, 139–148.

67. Patterson, W. J. Development of polymeric coatings for control of electro-osmotic flow in ASTP MA-011 electrophoresis technology experiment. NASA TMX-73311, National Aeronautics and Space Admin., Huntsville, AL, 1976.

68. Heard, D. H.; Seaman, G. V. F. Biochim. Biophys. Acta 1961, 53, 366–372.

69. Barlow, G. H.; Lazer, S. L.; Rueter, A.; Allen, R. In Bioprocessing in Space; NASA TM X-58191; Morrison, D. R., Ed.; L. B. Johnson Space Center: Houston, TX, January 1977; pp 125–132.

70. McGuire, J. K.; Snyder, R. S. In Electrophoresis '81, Allen, R. C.; Arnaud, P., Eds.; Walter deGruyter & Co.: Berlin, 1981; pp 947–960.

71. Hannig, K. In Modern Separation Methods of Macromolecules and Particles; Gerritsen, T., Ed.; Wiley Interscience: New York, NY, 1969; pp 45–69.

72. Saville, D. A. Physicochem. Hydrodyn. 1977 2, 893.

73. Biscans, B.; Alinat, P.; Bertrand, J.; Sanchez, V. Electrophoresis 1988, 9, 84–89.

74. McCreight, L. R. In Bioprocessing in Space; Morrison, D. R., Ed.; NASA TMX-58191, Lyndon B. Johnson Space Center: Houston, TX, January 1977, pp 143–158.

75. Strickler, A.; Sacks, T. Prep. Biochem. 1973, 3, 269–277.

76. Snyder, R. S.; Rhodes, P. H. In Frontiers in Bioprocessing; Sikdar, S. K.; Bier, M.; Todd, P., Eds.; CRC Press: Boca Raton, FL, 1989; pp 245–258.

77. Miller, T. Y.; Williams, G. P.; Snyder, R. S. Electrophoresis 1985, 6, 377.

78. Rhodes, P. H.; Snyder, R. S.; Roberts, G. O. J. Colloid Interface Sci. 1989, 129, 78.

79. Taylor, G. I. Proc. Roy. Soc. 1966, A291, 159–167.

80. Rhodes, P. H. In Electrophoresis '81; Allen, R. C.; Arnaud, P., Eds.; W. deGruyter & Co.: Berlin, 1981; pp 919–932.

81. Vanderhoff, J. W.; Micale, F. J.; Krumrine, P. H. In Electrokinetic Separation Methods; Righetti, P.G.; van Oss, C. J.; Vanderhoff, J. W., Eds.; Elsevier/North-Holland: Amsterdam, 1979; pp 121–141.

82. Ostrach, S. J. Chromatog. 1977, 140, 187.

83. Deiber, J. A.; D. A. Saville. In Materials Processing in the Reduced Gravity Environment of Space; Rindone, G. E., Ed.; North-Holland: New York, NY, 1982; pp 217–224.

84. Rhodes, P. H.; Snyder, R. S. In Materials Processing in the Reduced Gravity Environment of Space; Rindone, G. E., Ed.; North-Holland, New York, NY, 1982; pp 217–224.

85. Bauer, J.; Hannig, K. In Electrophoresis '86; Dunn, M. J., Ed.; VCH Verlagsgesellschaft: Berlin, 1986; pp 13–24.

86. Just, W. W.; Werner, G. In <u>Cell Separation Methods</u>; Bloemendal, H., Ed.; Elsevier/North–Holland Biomedical Press: Amsterdam, 1977; pp 131–142.
87. Just, W. W.; Werner, G. In <u>Electrokinetic Separation Methods</u>; Righetti, P. G.; Van Oss, C. J.; Vanderhoff, J. W., Eds.; Elsevier/North–Holland Biomedical Press: Amsterdam, 1979; pp 143–167.
88. Haydon, D. A.; Seaman, G. V. F. <u>Arch. Biochem. Biophys.</u> 1967, <u>122</u>, 126.
89. Levesque, M. J.; Nerem, R. M. <u>ASME J. Biomech. Engin.</u> 1985, <u>176</u>, 341–347. Stathopoulos, N. A.; Hellums, J. D. <u>Biotechnol. Bioengin.</u> 1985, <u>27</u>, 1021–1026.
90. Hannig, K.; Wirth, H. <u>Prog. Astronaut. Aeronaut.</u> 1977, <u>52</u>, 411–422.
91. Hannig K.; Bauer, J. <u>Adv. Space Res.</u> 1989, <u>9</u> (11), 91–96.
92. Mel, H. C. <u>J. Theoret. Biol.</u> 1964, <u>6</u>, 159–180.
93. Tippetts, R. D.; Mel, H. C.; Nichols, A. V. In <u>Chemical Engineering in Medicine and Biology</u>, Plenum Press, NY, 1967; pp 505–539.
94. Paulus, J. M.; Mel, H. C. <u>Exper. Cell Res.</u> 1967, <u>48</u>, 27–38.
95. Mel, H. C. <u>Chem. Engin. Science</u> 1964, <u>19</u>, 847–851.
96. Mel, H. C. In <u>Myeloproliferative Disorders of Animals and Man</u>; Clark, W. J.; Howard, E. B.; Hackett P.L., Eds.; U.S. Atomic Energy Commission: Washington, DC, 1970; pp 665–686.
97. Kolin, A. <u>J. Chromatogr.</u> 1967, <u>26</u>, 164–183.
98. Kolin, A. In <u>Electrokinetic Separation Methods</u>; Righetti, P. G; Van Oss, C. J.; Vanderhoff, J., Eds.; Elsevier/North–Holland Biomedical Press: Amsterdam, 1979; pp 169–220.

RECEIVED March 15, 1991

Chapter 16

Separation of Small-Cell Lung Cancer Cells from Bone Marrow Using Immunomagnetic Beads

F. J. Powers[1], C. A. Heath[1], E. D. Ball[2], J. Vredenburgh[2], and A. O. Converse[1]

[1]Thayer School of Engineering and [2]Dartmouth Medical School, Dartmouth College, Hanover, NH 03755

Immunomagnetic beads can be used to effect the removal of a subpopulation of cells from a mixed cell suspension in a flow-through system. One application of this process is the removal of tumor cells from bone marrow prior to its use in autologous bone marrow transplantation, a technique used to treat some forms of cancer. In our separator, the cell suspension flows through a 150 ml transfer pack which is held over an array of permanent magnets. Testing of the device on DMS-273 small cell lung cancer cells mixed with normal human bone marrow mononuclear cells resulted in a mean tumor cell removal of 3.64 logs (99.977%) with a concomitant mean normal cell colony forming unit (CFU) recovery of 61.3%. Direct (one antibody) and indirect (two antibody) methods of binding the beads to the cells were investigated. The extent of primary antibody coating of the beads was determined by flow cytometry. Also investigated were the effects of temperature, bead to cell ratio, and medium additives on tumor cell removal and normal cell recovery. The optimal conditions for separation were determined to be the indirect method of bead-cell binding at 22°C using a bead to tumor cell ratio of 25:1.

Autologous bone marrow transplantation (ABMT) is increasingly being investigated as a treatment modality for many forms of cancer, including small cell lung cancer, which comprises 25-30% of all the lung cancers (1). For ABMT, bone marrow is removed from the patient, after which the patient is treated with supralethal doses of chemotherapy, alone or in combination with radiation therapy. The marrow is subsequently given back to the patient in the bone marrow rescue. Unfortunately, in 60-80% of small cell lung cancer patients, the marrow is contaminated with tumor cells which must be removed before the bone marrow is transplanted (2). Current methods for bone marrow purging include the use of monoclonal antibodies (mAbs) and complement (3-6), cytotoxic drugs (7-9), immunotoxins (10,11), and immunomagnetic beads (12-21). Immunomagnetic beads are advantageous because they do not introduce biological products nor do they induce toxicity which occurs with complement and cytotoxic drugs, respectively. Also, for small cell lung cancer cells, immunomagnetic beads effect a better removal of tumor cells. Our objective was to design an immunomagnetic separation system capable of providing a three to five log tumor cell removal along with recovery of greater than 50% of normal bone marrow colony forming units.

Materials and Methods

Monoclonal Antibodies. Three tumor specific mAbs have been found to be the most effective combination of antibodies for removal of small cell lung cancer cells in previous experiments (Vredenburgh, J.J.; Ball, E.D. *Cancer Res.*, in press) and were used in this study. The mAbs are TFS-4 (IgG$_1$, 1 µg/10^6 tumor cells) from Medarex (W. Lebanon, NH), HNK-1 (IgM, 1 µg/10^6 tumor cells) from the mAb library, Dartmouth Medical School, and SCCL-175/95 (IgM, 10 µl ascites fluid/10^6 tumor cells). Thy-1 (IgM, 1 µg/10^6 tumor cells) from the mAb library, Dartmouth Medical School, and P-3 supernatant (IgG$_1$, 10 µl/10^6 tumor cells) from hybridomas were used as isotype-matched negative controls.

Immunobeads. The immunomagnetic beads used were sheep anti-mouse IgG coated and uncoated M-450 Dynabeads (Dynal Inc., Great Neck, NY). The beads have a diameter of 4.5 µm, a specific gravity of 1.5, and a magnetic susceptibility of approximately 10^{-2} cgs units.

Normal Cells. Normal mononuclear cells were isolated from bone marrow or peripheral blood, obtained from healthy, paid donors. MNCs were isolated using a ficoll-hypaque gradient technique (*22*).

Tumor Cells. DMS-273 small cell lung cancer cells (*23*) were cultured at 37°C in Weymouth's 752/1 medium (Gibco Labs, Grand Island, NY) containing L-glutamine (261 µg/ml, Gibco Labs), penicillin (89 units/ml, Gibco Labs), streptomycin (89 µg/ml, Gibco Labs), and 10% fetal calf serum (FCS, HyClone Labs, Logan, UT). The cells were harvested using enzyme-free cell dissociation solution (Specialty Media, Inc., Lavalette, NJ).

Medium for Separation Process. RPMI 1640 Medium (Gibco Labs) containing Hepes buffer (25 mM), L-glutamine (261 µg/ml), penicillin (89 units/ml), streptomycin (89 µg/ml), 0.5% human serum albumin (HSA) or 0.5% bovine serum albumin (BSA), and 0.005% DNAse was used in all separations.

Attachment of mAb to Cells. Normal and tumor cells were mixed (10% tumor cells and 90% normal MNC cells) and incubated with the mAbs for 60 minutes at 4°C on an Orbitron rotator (Boekel Industries, Inc., Model 260200) in 50 ml conical polypropylene tubes with a volume of 50 µl medium/10^6 cells. The total number of cells used varied as indicated in the following paragraphs but the stated cell concentrations were maintained. Following the incubation period, the cells were washed twice to remove any excess mAb which might interfere with the binding of beads to cells. The cells were pelleted between washings by centrifuging at 370g for 10 minutes (Beckman GPR Centrifuge). Finally, the cells were resuspended in 6.25 µl medium/10^6 cells.

Attachment of mAb to Beads for the Direct Binding Method. Uncoated M-450 Dynabeads were coated with one of the IgM mAbs, HNK-1 or Thy-1, by the procedure for physically adsorbing the mAbs directly onto the beads, outlined in the

Dynal product literature (#14001). Flow cytometry was used to determine the extent of mAb coating on the beads. Preparation of the beads (10^7) for flow cytometry consisted of two washes with phosphate buffered saline (PBS) containing 0.1% BSA, incubation on ice for 30 minutes with 25 μl FITC-conjugated goat anti-mouse mAb, a wash with PBS containing 0.1% BSA and 0.02% NaN_3, and resuspension in 200 μl of the PBS solution.

Attachment of Beads to Cells. In all experiments except those investigating the effect of bead to cell ratio, beads were added in a bead to tumor cell ratio of 25:1 at a concentration of $4x10^8$ beads/ml in a final volume of 12.5 μl/10^6 cells. The beads and cells were incubated for 30 minutes with the indirect binding method and for one hour with the direct binding method.

Separation Processes. Small scale separations (1-20x10^6 total cells) were used to determine the optimum conditions for large scale separations (>8x10^9 total cells). Small scale separations were carried out in 15 ml conical polypropylene tubes using a Dynal MPC-1 magnetic particle concentrator. The tube was placed in the separator for 1-2 minutes, during which time the beads and the cells attached to them were pulled to the side of the tube by the magnet contained in the separator. The supernatant containing the cells without attached beads was then collected by pipette.

Large scale separations were carried out using the flow-through magnetic separator illustrated in Figure 1. A 4x6 array of Neodymium-Iron-Boron permanent magnets, contained in a BioMag Separator obtained from Advanced Magnetics, Inc. (Cambridge, MA), was placed in a Lexan frame built to hold a 150 ml Fenwal transfer pack. The transfer pack was held over the magnet while the support allowed a uniform 2 mm thickness of fluid to pass through the bag. Ports were provided for attaching the inflow and outflow tubes to the transfer pack and a rubber gasket was used as a baffle to ensure that the bead and cell suspension passed over most of the magnets. A peristaltic pump (Millipore pump with Cole-Parmer cartridge pump head model 7519-00) was used to pump the suspension of beads and cells through the transfer pack in the separator to the collecting pack at a rate of 10 ml/minute. The system was then washed with approximately 120 ml medium to increase the recovery of normal cells and the effluent was collected in a separate 150 ml transfer pack.

Determining Tumor Cell Removal and Normal Cell Recovery. Since the DMS-273 small cell lung cancer cells and the normal MNCs are easily distinguishable under a microscope, hemacytometer counts were used to determine cell numbers both before and after the separation process. However, since it is also important to determine the effect of the separation process on the proliferating fraction of the cells, clonogenic assays were performed for both the tumor cells and the normal cells. Assays for tumor cells were performed by the limiting dilution technique. In a six well plate, tumor cells were plated in dilutions of 1:10 to 1:10^6. The tumor cell removal was calculated in the following manner:

$$\text{fraction removed} = \frac{\text{number of cells before separation}}{[(1/\text{dilution of last positive well})\text{x}(\text{plating efficiency})]}$$

Recovery of bone marrow colony forming units (CFUs), the cells which repopulate the marrow, was determined by methylcellulose cultures (24).

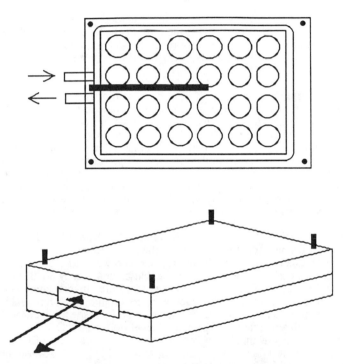

Figure 1. Large scale separator.   A 4x6 array of Neodymium-Iron-Boron permanent magnets, contained in a Lexan frame, retains the bead-bound cells and free magnetic beads as the suspension is pumped through the 150 ml transfer pack in the device. The support allows a uniform 2 mm thickness of fluid to pass through the bag. A rubber gasket (black rectangular object in the upper figure) is used as a baffle to ensure that the bead and cell suspension passes over most of the magnets.

Results

Direct v. Indirect Binding Methods. The indirect method refers to the attachment of beads coated with sheep anti-mouse antibody to tumor cells coated with the mouse mAb (the tumor specific mAb), while the direct method refers to the attachment of beads coated with the tumor specific mAb directly to the tumor cells. The two methods were compared in a series of experiments. Flow cytometry indicated that the beads used for the direct binding method were well coated with the mAb: 99.2 - 100% with respect to the Dynal M-450 Sheep anti-Mouse IgG Coated Beads.

As illustrated by Figure 2a, the indirect method was only slightly better than the direct method in terms of separating the tumor cells from the normal cells. Figure 2b, however, indicates a significant difference between the two methods in terms of nonspecific removal of tumor cells, with the direct method removing a greater percentage of cells nonspecifically.

Effect of Medium Additives. Different combinations of the medium additives heparin, BSA, and DNAse were tested on a small scale to determine which (if any) was causing a fibrous agglomeration of cells observed in a couple of preliminary large scale separation experiments. RPMI 1640 Medium containing Hepes buffer (25 mM), L-glutamine (261 μg/ml), penicillin (89 units/ml), and streptomycin (89 μg/ml) was used as the base medium and the effect of all possible combinations of heparin (1.0%), BSA (0.5%) and DNAse (0.006%) were investigated. The indirect separation process outlined above was used. Fibrous agglomeration of cells was observed only in the presence of heparin. Thus, heparin was not added to the medium in subsequent experiments.

Effect of Incubation Temperature During Attachment of Beads to Cells. In this series of small scale experiments, the incubation temperature during the attachment of beads to cells using the indirect method was varied since Hsu *et al.* (25) concluded that the conventional practice of incubation at 4°C may not be optimal for all cell types or separation systems. The four temperatures investigated were 0, 4, 22, and 37°C. The effect of the incubation temperature on tumor cell removal is shown in Figure 3. The plots show that both the specific and nonspecific removal of tumor cells increases with temperature, beginning to level off after 22°C.

The effect of this incubation temperature on the recovery of CFUs, the cells which repopulate the bone marrow, can be seen in Figure 4. Although the MNC recovery decreased with increasing temperature, the CFUs per MNC increased with temperature, so that the CFU recovery was constant over the temperature range.

Effect of Bead to Tumor Cell Ratio. In this series of small scale experiments, the effects of the bead to tumor cell ratios 1, 10, 25, and 50 to 1 on the separation process were investigated. The effect of bead to tumor cell ratio on the specific and nonspecific removal of tumor cells is illustrated in Figure 5a. This plot shows that both the specific and nonspecific tumor cell removal increase as the bead to cell ratio increases. The specific removal increases dramatically from ratios of 1:1 to 10:1, then more gradually as the ratio is increased above 10:1. Figure 5a also shows that the nonspecific tumor cell removal increases linearly with bead to tumor cell ratio in the range studied. Figure 5b shows that the normal cell recovery decreases as the bead to tumor cell ratio increases.

Large or Clinical Scale Separations. A number of large scale separations were performed using the flow-through separation system illustrated in Figure 1. The

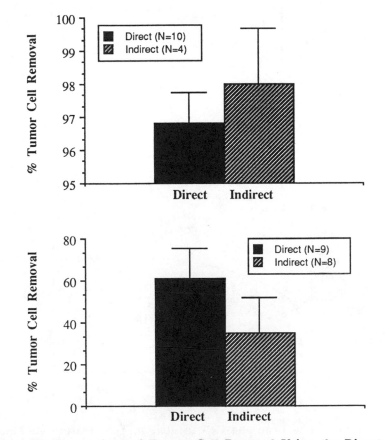

Figure 2. Comparison of Tumor Cell Removal Using the Direct and Indirect Binding Methods. (a) *Specific removal.* The antibody HNK-1 was used for both the direct (n = 10) and indirect (n = 4) binding methods. (b) *Nonspecific removal.* The antibody Thy-1 was used for the direct method (n = 9), while Thy-1 (n = 3) and P-3 (n = 5) were used for the indirect method. The bars represent 95% confidence limits.

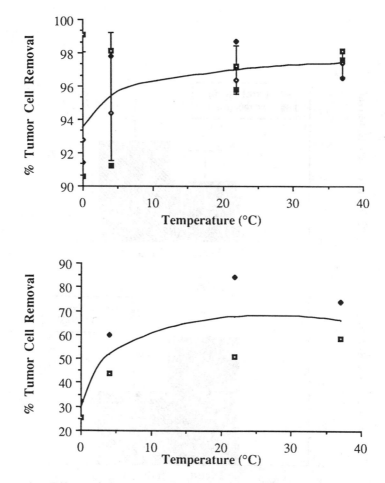

Figure 3. Effect of Bead and Cell Incubation Temperature on Tumor Cell Removal. (a) *Specific removal.* The tumor specific antibodies TFS-4 and HNK-1 were used (n = 4). (b) *Nonspecific removal.* The antibody P-3, an isotype matched negative control, was used (n = 2).

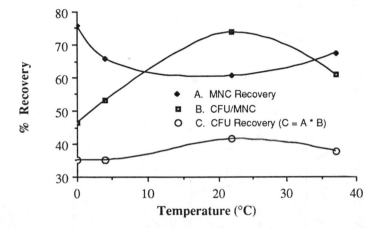

Figure 4. Effect of Bead and Cell Incubation Temperature on CFU Recovery. The separation process was carried out with only normal bone marrow mononuclear cells. The antibodies TFS-4, HNK-1, and SCCL-175-95 (n = 2) and P-3 (n = 1) were used and the data combined. CFU recovery was compared to that of untreated MNCs (n = 2).

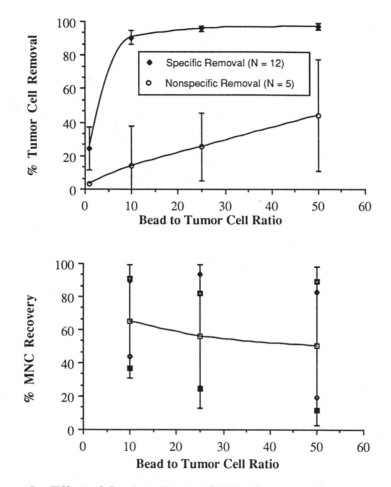

Figure 5. Effect of Bead to Tumor Cell Ratio. The bead to tumor cell ratios 1, 10, 25, and 50 to 1 were investigated. (a) *Tumor cell removal.* The tumor specific removal was done with the antibodies TFS-4, HNK-1, and SCCL-175/95, while the nonspecific removal was done with P-3. (b) *Normal MNC recovery.* The antibodies TFS-4, HNK-1, and SCCL-175/95 (n = 4) and P-3 (n = 1) were used.

results of these separations are summarized in Table I. The separation procedure resulted in a mean tumor cell removal of 3.64 logs (99.977%) over a 95% confidence interval of 3.01 to 4.27 logs (99.902 to 99.995%) with a mean normal cell CFU recovery of 61.3% over a 95% confidence interval of 27.9 to 94.7%.

TABLE I. Results of Large Scale Separations

Log Tumor Cell Removal	% MNC Recovery	% CFU-GM Recovery
2.64	73.7	nd[1]
3.00	72.3	nd
na[2]	82.9	63.1
3.22	29.0	nd
4.15	40.4	nd
4.40	58.0	nd
2.39	73.7	48.1
4.30	38.1	34.0
na	65.5	100.0
5.00	51.0	nd
Averages		
3.64	58.5	61.3
95% Confidence Intervals		
3.01 to 4.27	50.5 to 66.4	27.9 to 94.7

[1]nd: not determined

[2]na: not applicable (i.e., separations done on normal cells only)

Discussion

Small scale experiments with magnetic immunobeads and the mAbs TFS-4, HNK-1, and SCCL-175/95 were used to determine the optimal conditions for separating DMS-273 small cell lung cancer cells from normal bone marrow MNCs. A composition of 10% tumor cells was used throughout the experiments to simulate a worst-case scenario. Normal tumor cell contamination levels are 1% or lower. Investigation of the effect of different ratios of tumor cells to normals cells is planned.

The direct and indirect methods of bead cell binding were compared for efficacy. Each method has its advantages and disadvantages. The direct method uses only one antibody, effectively decreasing the cost of the procedure. However, the beads must be coated with the appropriate mAbs for each application, i.e, a separate batch of beads must be coated for each cancer type (and possibly for each patient). The indirect method uses two antibodies, possibly increasing cost, but the same beads can be used for each application; the antibodies coating the cells are the ones which need to be varied for different cancers or patients. Although the results varied considerably, the experiments demonstrated that the indirect method results in both a greater specific removal (although not significant) and a lower nonspecific removal of cells. The indirect method may overcome problems associated with the interaction between fixed antibodies on the bead and antigens on a mobile membrane (*26*). Future work includes an investigation comparing the effect of using different antibodies, alone and in different combinations, on the specific and nonspecific removal of cells.

The increase of both the specific and nonspecific removal of tumor cells with an increase in the incubation temperature, up to approximately 22°C, was also shown. The effect of this incubation temperature on the recovery of CFUs is shown in Figure 4. Although the MNC recovery decreased with increasing temperature, the CFUs per MNC increased with temperature, so that the CFU recovery was constant over the temperature range. Since a large portion of the procedure was done under a laminar flow hood at 22°C, it was most convenient to incubate the beads and cells at a temperature of 22°C.

Figure 5 shows, beyond a ratio of 25:1, that a small increase in tumor cell removal was accompanied by a large decrease in normal cell recovery. Based on these observations, the optimal bead to tumor cell ratio for separating small cell lung cancer cells from bone marrow was 25:1. The optimum bead to tumor cell ratio may be affected by the high percentage of tumor cells in the cell suspension. Lower tumor cell percentages, approaching physiological levels, may require a higher bead to tumor cell ratio as the tumor cells become more difficult to "find" and as a higher proportion of beads are lost to nonspecific adsorption.

Large scale separations of DMS-273 small cell lung cancer cells from normal bone marrow resulted in a mean tumor cell removal of 3.64 logs (99.977%) with a mean recovery of normal cell CFUs of 61.3%. The results of these experiments show that the separation of small cell lung cancer cells from bone marrow using immunomagnetic beads can be done on a clinical scale; clinical trials to treat patients with bone marrow purging using this method are planned for the near future.

Investigations to improve the process by increasing both normal cell recovery and tumor cell removal are underway. For example, the loss of CFUs may be reduced as a result of coating the beads with lipids as done by Kedar et al. (27). Tumor cell removal can often be increased by treating the cells with two cycles of purging instead of one, yet at the expense of reduced normal cell recovery. Other modifications are also being considered.

Acknowledgments

The authors thank Medarex, Inc. for generously supplying the TFS-4 mAb used in this study.

Literature Cited

1. Gazdar, A.F.; Linnoila, I. Semin. Oncol. 1988, 15, 215.
2. Stahel, R.A.; Mabry, M.; Skarin, A.T.; Speak, J.; Bernal, S.D. J. Clin. Oncol. 1985, 3, 455.
3. Jansen, J.; Falkenburg, J.H.F.; Stepan, D.E.; Le Bien, T. Semin. Hematol. 1984, 21, 164.
4. Bast, R.C.; DeFabritiis, P.; Lipton, J.; Gelber, R.; Maver, C.; Nadler, L.; Sallan, S.; Ritz, J. Cancer Res. 1985, 45, 499.
5. Ramsay, N.; LeBien, T.; Nesbit, M.; McGlave, P.; Weisdorf, D.; Kenyon, P.; Hurd, D.; Goldman, A.; Kim, T.; Kersey, J. Blood 1985, 66, 508.
6. Ball, E.D.; Mills, L.E.; Coughlin, C.T.; Beck, J.R.; Cornwell, G.G. Blood 1986, 68, 1311.
7. Sharkis, S.J.; Santos, G.W.; Colvin, M. Blood 1980, 55, 521.
8. Korbling, M.; Hess, A.D.; Tutschka, P.J.; Kaizer, H.; Colvin, M.O.; Santos, G.W. Br. J. Haematol. 1982, 52, 89.
9. Reynolds, C.P.; Reynolds, D.A.; Frenkel, E.P.; Smith, R.G. Cancer Res. 1982, 42, 1331.

10. Muirhead, M.; Martin, P.J.; Torokstorb, B.; Uhr, J.W.; Vitetta, E.S. *Blood* **1983**, *62*, 327.
11. Colombatti, M.; Colombatti, A.; Blythman, H.E. *J. Natl. Cancer Inst.* **1984**, *72*, 1095.
12. Dicke, K.A. In *Recent Advances in Bone Marrow Transplantation*; Gale, R.P., Ed.; Alan R. Liss: New York, NY, 1983, pp 689-702.
13. Treleaven, J.G.; Gibson, F.M.; Ugelstad, J.; Rembaum, A.; Philip, T.; Caine, C.D. *Lancet* **1984**, *1*, 70.
14. Kemshead, J.T.; Treleaven, J.G.; Gibson, F.M.; Ugelstad, J.; Rembaum, A.; Philip, T. In *Advances in Neuroblastoma Research*; Evans, A.E., D'Angio, G., Seeger, R.C., Eds.; Alan R. Liss: New York, NY, 1985, pp 413-424.
15. Kemshead, J.T.; Heath, L.; Gibson, F.M.; Katz, F.; Richmond, F.; Treleaven, J.; Ugelstad, J. *Br. J. Cancer* **1986**, *54*, 771.
16. Reynolds, C.P.; Moss, T.J.; Seeger, R.C.; Black, A.T.; Woody, J.N. In *Advances in Neuroblastoma Research*; Evans, A.E., D'Angio, G., Seeger, R.C., Eds.; Alan R. Liss: New York, NY, 1985, pp 425-442.
17. Reynolds, C.P.; Biedler, J.L.; Spengler, B.A.; Reynolds, D.A.; Ross, R.A.; Frenkel, E.P.; Smith, R.G. *J. Natl. Cancer Inst.* **1986**, *76*, 375.
18. Reading, C.L.; Takaue, Y. *Biochim. Biophys. Acta* **1986**, *865*, 141.
19. Kvalheim, G.; Fodstad, O.; Pihl, A.; Nustad, K.; Pharo, A.; Ugelstad, J.; Funderud, S. *Cancer Res.* **1987**, *47*, 846.
20. Kvalheim, G.; Sorenson, O.; Bodstad, O.; Funderud, S.; Diesel, S.; Dorken, B.; Nustad, K.; Jakobsen, E.; Ugelstad, J.; Pihl, A. *Bone Marrow Trans.* **1988**, *3*, 31.
21. Nustad, K.; Michaelsen, T.E.; Kierulf, B.; Fjeld, J.G.; Kvalheim, G.; Pihl, A.; Ugelstad, J.; Funderud, S. *Bone Marrow Trans.* **1987**, *2(Suppl. 2)*, 81.
22. Boyum, A. *Scand. J. Immunol.* **1976**, *5*, 9.
23. Pettengill, O.S.; Sorenson, G.D. In *Small Cell Lung Cancer*; Greco, Oldham, Bunn, Eds.; Grune and Stratton: New York, NY, 1981, pp 51-77.
24. Howell, A.; Ball, E.D. *Blood* **1985**, *66*, 649.
25. Hsu, S.C.; Yeh, M.M.; Chu, I.M.; Chen, P.M. *Biotechnol. Tech.* **1989**, *3*, 257.
26. Kemshead, J.T.; Elsom, G.; Patel, K. In *Bone Marrow Purging and Processing*; Gross, S., Gee, A.P., Worthington-White, D.A., Eds.; Alan R. Liss: New York, NY, 1990, pp 235-251.
27. Kedar, A.; Schreier, H.; Rios, A.; Janssen, W.; Gross, S. In *Bone Marrow Purging and Processing*; Gross, S., Gee, A.P., Worthington-White, D.A., Eds.; Alan R. Liss, Inc.: New York, NY, 1990, pp 293-301.

RECEIVED March 15, 1991

Chapter 17

Analytical- and Process-Scale Cell Separation with Bioreceptor Ferrofluids and High-Gradient Magnetic Separation

P. A. Liberti and B. P. Feeley

Immunicon Corporation, 1310 Masons Mill II, Huntingdon Valley, PA 19006

The development of small colloidal magnetic particles (< 0.1u) and their application to cell separations via high gradient magnetic methods is reviewed. An analysis of the properties of these particles contrasted to large magnetic particles (1-5u) argues for the superiority of the former even though their use has been plagued by non-specific binding issues and the use of crude high gradient magnetic separators. The development of a BioReceptor (antibody, enzyme, lectin) coated Ferrofluid and two novel high gradient magnetic separators are described. Both separators can be used for analytical scale cell separations (200 - 300 ul volumes) and one of them can easily be adapted for process scale separations. Performance data on separations done in simple and complex matrices is given.

The most discriminating means for separation is based on selection via complimentarity at the molecular level as for example via receptor-ligand interaction. The availability of bio-receptor molecules such as monoclonal antibodies (MAB) or lectins that have affinity for well defined surface molecular structures of target macromolecules is the basis for the broad use of affinity separations in modern biology. For the isolation of macromolecules, affinity techniques are relatively straightforward as bio-receptors can readily be covalently attached to a variety of solid supports such that removal of target molecules from solution is a simple task. Subsequent recovery of target product generally involves dissociation of receptor-ligand interaction via mild denaturation of these components or employs competing ligand.

The extension of affinity techniques from molecule isolations to cells, involves substantially more complex issues because of:

(1) the complexity of cell membrane structures and the heterogeneous chemical nature of such structures,

(2) the mobility of intrinsic membrane macromolecules in the fluid lipid bilayer as well as normal shedding of such macromolecules,

0097–6156/91/0464–0268$06.25/0

(3) the size of cells compared with macromolecules (1000 to 2000 times greater),

(4) the complications arising from cell death and the spilling of complex cell components into the system,

(5) the complications introduced by normal cell function and in particular, functions triggered by the binding of cell surface receptors by macromolecules and,

(6) the difficulty of reversing receptor-ligand interaction, particularly those involving MAB's, while maintaining cell integrity.

Based on these considerations, initial cell affinity separations were confined to fractionation of cells on adsorbant surfaces (Petri dishes) or fibers packed into columns where bio-receptors had been covalently coupled or absorbed onto such surfaces (*18*). Although such approaches have proved useful in the laboratory, they are difficult to perform and to reproduce and, further, they do not scale up well.

Kemshead and Ugelstad, who pioneered the use of magnetic particles and MAB's to cell surface components for affinity cell separations (the Dynal system), were led to magnetic retrieval after having considered the fundamentals of cell separation and after having evaluated fibers as well as other chromatographic supports (*19*).   By using magnetic particles, the difficulties in passaging cells through affinity supports such as entrapping of cells, clogging and shearing effects were obviated with the additional advantage of having receptors on all sides of a cell as attachment points as compared with those tangential to one surface. Thus by appropriate attachment of magnetic particles cells can literally be "pulled" and "pushed" through solution.

This manuscript presents a review of work done on cell separation with colloidal sized magnetic particles.   It further describes work done in our laboratory on the development of protein coated ferrofluids, analytical scale cell separations with those materials and a new means for generating the kinds of magnetic gradients required for separating cells labelled with colloid magnetic materials.

Magnetic Materials For Cell Separation

The development of useful magnetic particle technology in separations has been made possible by the use of superparamag-netic particles such as magnetite, Fe_3O_4. Superparamagnetic materials result when the dimensions of a crystal of ferromagnetic material are smaller than about 300 A, which is the approximate size of a magnetic domain, i.e. the minimum volume element capable of retaining a permanent magnetic dipole (*1*).   Such crystals at or below that size become magnetic dipoles when placed in a magnetic field but lose their magnetism when the field is turned off.   Hence, individual crystals or particles composed of assemblies of such crystals can be readily resuspended after the application of a magnetic field because they retain no permanent magnetic dipole and accordingly are not attracted to each other.

Ferrofluids, first described by Elmore in 1938 (*2*), are colloidal solutions containing superparamagnetic crystals of magnetite coated with surfactants such as sodium lauryl sulfate. They can be prepared in sizes ranging from 5 to 60 nanometers, are true lyophilic colloids and have the thermodynamic properties of solutions (4). The discovery by Molday and MacKenzie (11) that ferrofluids coated with dextran could be prepared by forming magnetite (treatment of ferrous and ferric chlorides at 1:2 molar ratio with base) in the presence of 25% aqueous dextran with subsequent removal of product by gel filtration led to the development of a particle to which antibodies could be coupled via conventional cyanogen bromide chemistry.

Another type of coated superparamagnetic colloid used for cell separation has been developed by Yau and co-workers (22). They form a cobalt metal colloid by reduction of dilute cobalt citrate with $NaBH_4$. The metal colloid is peptized with NaOH, avidin is adsorbed and next mildly succinylated with succinic anhydride. Purification with gel chromatography produces a useful product which must, however, be used within 24 hours as the colloid rapidly oxidizes.

In 1984 Owen discovered a simple co-precipitation method for preparing a bovine serum albumin (BSA) coated ferrofluid to which specific antibodies could be coupled employing heterobifunctional cross-linking agents (14-15). The synthesis of the BSA ferrofluid involves mixing aqueous solutions of protein with ferrous and ferric chlorides (2 moles Fe^{+3}: 1 mole Fe^{+2}) and addition of NH_4OH. After washing, the resultant co-precipitate is resuspended into the colloidal or ferrofluid state with low ionic strength buffers and occasionally with mild sonication. This work was extended by Liberti and co-workers who discovered that if dilute base is used in the above process and that if the base is added such that the pH of the mixture remains below 9.0, proteins such as antibodies can be coated onto ferrofluids in a manner where the protein retains biological activity (16).

An alternative procedure resulting in nearly identical materials involves direct coating via sonication of polymer or protein solutions in the presence of appropriately treated bare crystalline magnetite (7). For this procedure , just as with the modified co-precipitation method, when a bio-receptor such as MAB, lectin or enzyme is incorporated into the coating protein material, biological activity and specificity of immobilized bio-receptor is retained. In the case of MAB's, as much as 40% of theoretical binding capacity is retained. Further, for the immobilization of nearly 25 different monoclonal and polyclonal antibodies no changes in specificities have been observed. For both the co-precipitation and sonication procedures for synthesizing bio-receptor ferrofluid, protein or polymer binds to magnetite crystals via electrostatic, dipole-dipole and coordinate bonding. These interactions are generally sufficient to prevent leaching of these materials from the magnetite crystal. Occasionally cross-linking agents have been employed to perform intraparticle cross-linking to give greater stability. To date protein ferrofluids prepared by these methods have been employed for performing nearly twenty immunoassays.

Large magnetic particles (4.5 micron) used in cell affinity separations have been synthesized by polymerizing polystyrene (latex) (20) or polyacrolein in the presence of ferrofluid (8). As the growing polymer "winds" to form spheres, superparamagnetic crystals of the ferrofluid are trapped in the interstices of the growing particle. By appropriate chemical modification of the surfaces of such particles, antibodies can be covalently attached and particle surface character can be obtained such that minimal non-specific interaction with cell membranes occurs.   Another large superparamagnetic particle which has been used in cell separations is also available through Advanced Magnetics, Cambridge, MA (*21*). This particle, which is considerably heterogeneous in size, is an agglomerate produced by treating ferrofluid derivative with polymer silane and subsequently introducing amino groups for the chemical coupling to antibodies and other proteins.

Large Versus Small Magnetic Particles

Assuming for the moment that retrieval of cells affinity labelled with magnetic material, large or small, can be accomplished with equal ease, it is instructive to consider the relative merits of large particles versus ferrofluid. Some relevant issues are listed in Table I.   The more superior particle as it relates to a given property for cell separations is indicated as such.   Regarding the kinetics of labelling cells, the ferrofluids are clearly superior as they react with diffusion limited rates (chemically they are solutes) and require no mixing compared with large particles which require mixing to cause collisions. For labelling reactions which require 30 minutes with Dynal particles those employing BioReceptor Ferrofluid can be done in 5-10 minutes.   On positive selection, i.e removal of the population of cells of interest for subsequent use, large particles typically form cages of magnetic material around positively selected cells and the removal of the particles from the cells requires culturing (usually 24 hours - and even then is not particularly effective).   On the other hand, for cells isolated in ferrofluid there is no visible indication of material on the cell membrane and as such, various manipulations can generally be undertaken immediately following isolation.   The relationship between non-specific binding and particle size should theoretically be a function of particle size or the contact area between the binding particle and the cell surface.   The basis for this statement is that if a small and a large binding sphere are composed of the same surface material then non-specific interaction must be a sum of individual interactions and accordingly related to the potential area of contact.   Another consideration is that the ferrofluids have diffusive energy and consequently, following a non-productive collision with a cell surface, ferrofluids have the property of diffusing back into the solution.   This, coupled with the fact that large particles gravitationally settle onto cells, theoretically argues that smaller is considerably better.   In practice it has been found that dextran coated ferrofluids, for reasons which are unclear, have substantial non-specific binding to cell surfaces.   On the other hand, appropriately coated protein ferrofluids show low non-specific binding.

Regarding the potential of dislodging a labelled cell receptor and since cell surface receptors are indeed shed and undoubtedly can be dislodged from the membrane by external forces, then it would seem that a labelling particle which can bind to multiple receptors should be superior to a particle of lessor capacity.   In that category, if dislodgement is a significant issue, then larger particles where one particle can make many receptor contacts would have an advantage.

Magnetic Separation

The magnetic force which a cell labelled with superparamagnetic materials experiences is a function of two variables: the magnetic moment which the external field can induce on the labelling material and the gradient of the magnetic field at the location of the cell. For large labelling particles which, as noted, are assemblies of ferrofluid particles, the magnetic moment is large and hence the field gradient required to move a cell through the medium need not be large. On the other hand, cells labelled with ferrofluids require substantially higher gradients because these smaller particles will be less magnetic.

High gradient magnetic separation (HGMS) is a well established industrial process for removing weakly magnetic materials from complex mixtures such as red iron oxides from clay slurries to produce kaolin used to size writing paper (3, 5, 13). The simplist design for high gradient magnetic separation is to place magnetic stainless steel wool into a tube which is then placed between the poles of a magnet. The physics of generating a high gradient magnetic field on a single stainless steel filament is shown in Figure 1. The gradient field produced around the filament is a resultant of the superposition of the external field and of the field induced on the filament. Note that field line cancellation occurs on the sides of the filament because the direction of the field lines of the induced field are opposite to those of the external field. Hence, a gradient field is established around the filament forcing magnetic materials to the sides of the filament facing the external poles. From Maxwell's equations it can be shown that an inverse relationship exists between the diameter of the filament and the gradient of the field induced. Hence, the smaller the filament diameter the greater the field gradient. In such arrangements it is possible to generate field gradients of 100 - 200 Kilo Oesterds per centimeter from external fields of 10 - 15 Kilo Oesterds.

The experimental apparatus employed to perform high gradient magnetic separation on cells have been laboratory variants of industrial separators (9, 12, 14, 15, 17, 22). Briefly, small glass columns (around 10 cm x 0.5-1 cm i.d.) loosely packed with fine magnetic grade stainless steel wool (approximately 25 micron diameter) have been placed between the poles of electromagnets which can generate 10 - 15 Kilogauss (kG) fields. The columns have been fitted with appropriate inlet and outlet valves. Magnetically labelled cells are pumped through the column, labelled cells are retained on the steel wool, next the field is removed and cells are retrieved by flow and usually by gentle vibration of the column. A variant of these systems is currently commercially available and is called the Macs Cell Sorter (10).

In preliminary experiments with magnetic arrangements similar to those described above, significant difficulty was experienced in reproducing results and further it was found that great care and special, considerably elaborate techniques had to be employed to prevent non-specific trapping of cells. Additionally, it was found that due to the size of these columns, significant quantities of experimental materials were required. This becomes a barrier to performing various useful laboratory separations. Thus the design of a high gradient separator which could function in a microtitre well arrangement so as to be consistent with volumes and protocols normally used by cellular experimentalists seemed a worthwhile goal. From basic studies on high gradient

Table I. Relative Merits of Large (> 1 micron)
Versus Small (<0.06 micron) Magnetic
Particles For Cell Separation

Property	Large	Small
Kinetics of Labelling		Superior
Positive Selection		Superior
Non-Specific Binding		Superior
Negative Potential For Receptor Dislodgement	Superior	
Settling/Solution Stability		Superior
Surface Area		Superior

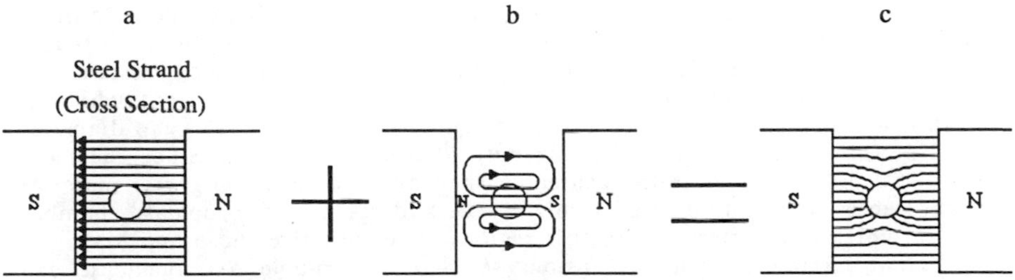

Figure 1. Principle of Establishing a High Gradient Magnetic Field.
Traditionally high gradient magnetic fields have been established by
placing ferromagnetic materials such as steel wool in a uniform magnetic
field. The origin of the gradient field around one such steel wool strand
results from the superposition of the external field and the field induced
on the strand and can be understood by reference to the figures where a.
shows the field line vectors (flux density) for the external field and note
the direction (North to South) and b. shows the induced field on the steel
strand. Note that the induced field gives rise to a bar magnet cross
sectionally in the strand and that the poles of this magnet are North on
the left of the strand and South on the right, i.e. opposite polarity of the
external field. As the field vectors of the induced field and external field
are of opposite direction, they cancel giving the sum shown in c. The high
density of flux lines on the sides of the strand facing the external field
poles relative to the density above and below the strand gives rise to the
high gradient. Thus magnetic material is drawn to those sides of the
strand facing the poles.
a. Steel strand in the external magnetic field.
b. Magnetic field generated by steel strand.
c. Sum of a. and b.

separation of ferrofluid, the surprising discovery was made that high gradient fields which are sufficient to separate labelled cells, can be established on iron wires, one to two orders of magnitude greater in diameter than has been traditionally used, i.e. diameters of 0.8 to 3mm. This discovery enabled the fabrication of ferro-magnetic structures for doing high gradient magnetic separations which do not deform when placed into or removed from strong magnetic fields and to which ferrofluid labelled cells would be attracted via the magnetic gradient established on such structures.

Another innovation these basic studies led to was the extension of the concept of open field gradients to the use of quadrupole (2N pole) fields for separating ferrofluids or appropriately labelled cells. Open field gradients exist at the corners of magnetic pole pieces or can be achieved between poles of a magnet by shaping pole pieces appropriately. For example, when a flat pole piece is opposed by a triangular shaped pole, the field lines converge to the apex of the latter and thus a gradient of flux lines (gradient field) is created. Kronick (6) and Molday and MacKenzie (11) have used such fields to collect labelled cells to the sides of a vessel. The use of such field gradient generating arrangements is limited because only relatively small field gradients can be achieved. Quadrupoles or multipole arrangements, on the other hand, are essentially unlimited as to the gradient which can be established. This can be seen by reference to Figure 2A. which depicts a typical quadrupole arrangement with North poles opposed and South poles opposed. Note that at the mid-point between two like poles, the field is 0. If field lines are superimposed on Figure 2A by allowing flux lines to flow from one pole piece to its neighbors on either side (Figure 2B.) open field gradients directed radially to the pole pieces result. By appropriate pole piece design and placement, radial gradient fields can be established. The gradient which is established in such an arrangement is determined by one-half the gap diameter (H=0 at the center) and the field strength at the poles' faces. Since the field and field gradient is 0 at the center of such arrangements, magnetic material at the center will feel no force. This can be addressed by appropriate chamber design which eliminates cells in this region but in practice there appears to be sufficient mixing in cell suspensions so that quadrupole arrangements can effectively pull labelled cells to the sides of cylindrical tubes.

Cell Separation Devices

As already noted a key goal for developing high gradient magnetic separation devices for magnetic cell sorting was to find a means to eliminate the use of the steel wool matrix previously employed. Although this form of matrix has proven to be a valuable tool for high gradient purification of ferrofluid as well as having a proven utility in the mining industry, this form of matrix can be harmful to cells and results in higher non-specific removal of cells than methods which might eliminate using steel wool. Based on the discoveries noted above, two classes of high gradient magnetic sorters which offer significant advantages over existing devices were developed. The first which generates gradient fields with an internal ferromagnetic structure was designed for cell sorting applications which can be performed in a manual or semi-automated fashion in microtitre plate wells. The second which incorporates the quadrupole concept can be utilized to construct a system for an appropriately designed carrousel-type microtiter plate which could be easily automated or utilized for batch separation, see below, or for continuous operation.

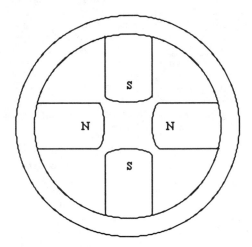

Figure 2.   A. Typical pole configuration in quadrupole magnet.

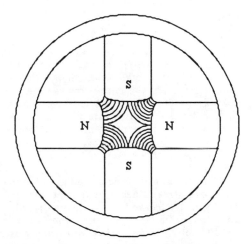

Figure 2.   B. Radial gradient flux density created by quadrupole magnetic arrangement.

Figure 3 shows several views of the internally generated gradient system developed for microtitre well use as well as a depiction of its placement in a permanent magnet designed for it. The device consists of a non-magnetic T-shaped frame into which 2 ferromagnetic wires, each bent in a loop shape are inserted for each separation to be performed. The bottom section of the T-shaped frame allows the device to be anchored between two pole faces of the magnet whereupon a high gradient field is induced on the wire loops. The sets of wire loops which extend downward from the frame are spaced apart such that when raising a strip of microtitre wells upward (via a fabricated elevation device which fits onto the yoke) each well has a pair of wire loops within it. Each wire is composed of ferromagnetic galvanized steel, is 0.8mm in diameter and when looped has approximately 1.6cm of usable capture length. The two looped wires per set are separated by 1.6 mm.

In a typical cell separation experiment, cell suspension, previously labelled with a specific ferrofluid reagent, is pipetted into microtitre wells. The microtitre strip is positioned in a holder and next into the space under the T-shaped loop holder in the permanent magnet device. Employing a simple mechanical device the suspensions are raised up onto the wires whereupon labelled cells experience the magnetic force generated by the gradient on the loops. Cells specifically labelled with the magnetic reagent are pulled onto the wires and thus immobilized within minutes. The cups can then be lowered and subsequent wash steps can be performed on the immobilized cells (still in field) using fresh buffer. If desired, the T-shaped loop holder with cells immobilized on the wires can then be removed from the field and cells eluted from the wire loops into fresh buffer. The T-shaped devices can be manufactured inexpensively such that they can be disposed of or could be autoclaved if desired for reuse.

A scheme used successfully to determine specific cell depletion is to label all cells with [51]chromium prior to the depletion experiment and determine the total counts going into each well. Separation is performed and by using snap apart microtitre wells the counts remaining in the supernatant (cells not separated) and wash buffer can be determined. These counts are then subtracted from the total counts. Knowing this number and the exact counts per cell one can easily calculate the percentage of cells removed.

Some of the key advantages this system offers are:

(1)   The wires are arranged perpendicular to the magnetic field maximizing gradient field, are streamlined and do not overlap or contact each other resulting in zero physical entrapment. Additionally, the loops are rigid and do not deform in the magnetic field. Further, they contain ample surface area to capture over 2.5×10^5 cells per microtitre well (volumes of 250-300ul) in a monolayer fashion. In many applications this arrangement eliminates the need for a wash step.

(2)   The separation is performed under near zero-flow conditions thus eliminating the loss of immobilized cells occuring in systems where shearing takes place from the flow of reaction mixtures or wash buffer.

(3)   Separations are performed in standard microtitre plate wells which makes this device useful in a wide range of cell sorting applications.

1. Cross-Section (To Scale)

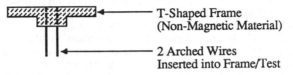

T-Shaped Frame
(Non-Magnetic Material)

2 Arched Wires
Inserted into Frame/Test

2. Side View

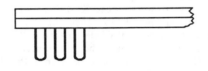

3. 3/4 View

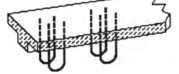

4. Device Shown Inside An HGMS Magnet

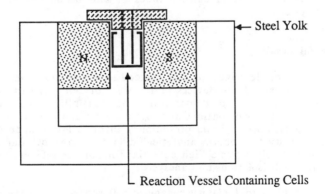

Steel Yolk

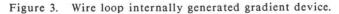

Reaction Vessel Containing Cells

Figure 3.   Wire loop internally generated gradient device.

The quadrupole cell separation device is constructed by merely placing an appropriate test tube or cylindrical flow-through tube in a quadrupole magnetic arrangement as already shown in Figure 2. This arrangement performs cell separations with no internal wires; instead cells labelled with ferrofluid are pulled to and collected on the walls of the reaction vessel via the externally generated radial gradients.

The scheme used to date for testing and optimizing the quadrupole system is to perform batch separations simply by placing a reaction vessel such as a 15 ml cell culture tube (which fits snugly between the 4 magnets) in the quadrupole device. Since the field drops to zero in the very center and no separation takes place in that region it is convenient to place a hollow plastic tube which runs vertically down the center of the reaction vessel which serves well for the introduction of reaction mixtures, drawing off of cell supernatants after separation, introduction of wash buffer (if needed) as well as for the elution of positively selected cells. This system can be fabricated in a variety of sizes for batch separations as described here or could be done continuously with an appropriately designed flow-through cell. As noted above, it can be adapted for an automated, multiple reaction format where a carrousel with individual reaction wells could be mechanically manipulated into and out of the magnetic field. It should be noted that this latter scheme would be possible with a quadrupole arrangement or with a multipole arrangement.

Some of the advantages of this cell separation device are:

(1)   no internal wires or gradient material is required which will greatly simplify automation and sterilization of components.

(2)   the system has the versatility of adaptation to singular manual separations in conventional culture tubes or can be developed into a fully automated system.

Experimental Results

This section describes some of the cell separation experiments that have been performed using these devices in a variety of cell matrices. The goal for these experiments was to demonstrate the feasibility of doing affinity separations on cell subpopulations and to determine how well these systems can be used for enrichment. From preliminary information it appears that these systems can be used to obtain analytical data for many cellular applications. Further, the technology has a high potential for further refinement and should extend to rare cell isolations and analysis.

One problem that plagues magnetic cell separation systems is the non-specific binding (NSB) of particles to cells. Therefore an immediate goal was to identify the degree of NSB using a goat-anti-mouse Fc coated ferrofluid and to reduce this to an acceptable level for feasibility studies. This was accomplished by formulating a variety of cell compatible buffers and testing the non-specific interactions between cells and ferrofluid particles.

Tests for non-specific interaction were done visually on a microscope by observing cells which had bound ferrofluid move to the wire in a gradient field. A convenient and simple arrangement for doing the former is to place a single taut wire on a microscope slide and affix the ends with epoxy. By placing a single bar magnet near and parallel to the wire a droplet of cell suspension can be observed and the movement of cells therein viewed under the influence of the field induced on the wire.

Using such an arrangement non-specific binding to various cell types i.e. T and B lymphocytes, monocytes, granulocytes and red blood cells by various ferrofluid preparations could rapidly be tested. By varying formulations of the latter it became possible to achieve levels of NSB to 0.5% or less.

Labelling and Depletion of Cultured Human T-Cells. To test the wire loop device and a goat anti-mouse Fc ferrofluid [GAMFcFF], the depletion of cultured human T-cells [ATCC CCL119 CCRF-CEM] employing anti-CD45 MAB was examined as a function of incubation time. In an immunofluorescence study using flow cytometry, this cell line showed positive expression of both CD45 and CD4 antigens. [The CD45 MAB is a common leukocyte marker while CD4 MAB is a T-helper cell marker.] In all cases MAB is preincubated with GAMFcFF which captures the Fc portion of the mouse MAB. In this fashion one can use this as a single reagent for cell isolations. The level of MAB incubated with ferrofluid is dependent on the amount of goat-anti-mouse Fc on the particle and the final single reagent (ferrofluid plus MAB) is always formulated such that no excess MAB is present. When this reagent is maintained at 2-8°C, it has excellent long term stability. In a typical depletion experiment cells are first labelled with ^{51}chromium for 1 hour at 37°C (cells at $1x10^7$ cells/mL) washed 2x in PBS and resuspended to $2x10^6$ cells/mL (c/mL) in a cell compatible buffer (CB3). Cells are then labelled with ferrofluid particles previously diluted to a working concentration in ferrofluid buffer (CB3c) and incubated with the MAB to be used for the particular study. The Bio-Receptor Ferrofluid/cell reaction will be referred to as the "cell labelling reaction". Note also that all results reported are minimally for duplicate analysis.

Typically, per reaction, ^{51}chromium labelled cells at $2x10^6$ c/mL are combined 1:1 with the diluted, MAB labelled ferrofluid. Volumes can be varied depending on the concentration of cells desired and the ferrofluid concentration needed for the separation. In this particular experiment 125 ul of diluted ferrofluid/anti-CD45 reagent (1:50 in CB3c) was combined with 125 ul ^{51}chromium labelled T-cells at $2x10^6$ c/mL in CB3 and incubated for either 15 or 30 minutes at room temperature. Controls included a case where no ferrofluid was used (cells only) and a case where ferrofluid was used with cells and no anti-CD45. Cells were separated for 5 minutes in an 8.25 kG permanent magnet using the wire loop device in Figure 3. Depletion was determined as described above in the Cell Separation Devices section.

The results for this experiment are shown in Table II and demonstrate that labelling is complete by 15 minutes. Also note that non-specific binding values (15 minute reaction) as calculated from supernatant analysis on cells plus ferrofluid and cells minus ferrofluid are 5.6 and 5.0 respectively. From results of experiments (not shown) using higher concentrations of cells and higher levels of free ^{51}chromium it was found that the 5% NSB is primarily due to material left on the wire loops and is primarily free ^{51}chromium.

Therefore the real % NSB in this particular system is between 0.5 and 1.5%, i.e. the difference between the non-specific controls.   The standard deviation around the % depletion is +/- 0.75 for all experiments reported herein.

The effect of field strength on the specific depletion reaction was examined by performing depletions with the T-shaped wire loop arrangement placed in an electromagnet where field strength could be varied.   The microtitre wells were manipulated into and out of the field with the aid of a fabricated elevation device.   Reaction mixtures consisted of 125 ul of diluted ferrofluid/anti-CD45 reagent (1:50 in CB3c) + 125 ul ^{51}chromium labelled T-cells at 2.4 x 10^6 c/mL in CB3 and reacted for 15 minutes at room temperature. All mixtures were separated for 5 minutes at the various field strengths. Depletion was determinsted as above. Figure 4 depicts these results and shows that above 6.3 kG no further depletion takes places. As permanent magnet fixtures of 6-9 kG are readily fabricated there is no need for higher field electromagnets to perform these separations. Currently fixtures which generate 7 - 7.5 kG in an air gap of 0.95cm are used.

Next, a study was performed to examine percent depletion as a function of high gradient magnetic separation time.   The reaction mixtures were identical to that of the previous study.   Field strength was constant at 7 kG. Figure 5 depicts the results of this study.   As can be seen from the figure, depletion plateaus by 3 minutes but, depending on the application, intervals of less than 1 minute might suffice.

To   determine   the   effect   of   labelling   reagent   concentration [GAMFcFF/anti-CD45] on specific and non-specific depletion, dilutions of reagent over nearly a ten fold range were used to label and deplete cultured T-cells. Goat anti-mouse Fc ferrofluid was first diluted 1:5, 1:10, 1:20 and 1:40 in CB3c and incubated with anti-CD45 MAB. [As the ferrofluid dilution increased from 1:5 to 1:40 the level of anti-CD45 was dropped proportionally to avoid excess MAB conditions.]   125 ul of the diluted ferrofluid/anti-CD45 reagent was then combined with 125 ul ^{51}chromium labelled T-cells at $4x10^6$ c/mL in CB3. In order to assess non-specific binding, controls were also set up wherein diluted ferrofluid was used without anti-CD45 MAB. These results are shown in Figure 6.   Depletion percentages increase with increasing reagent concentration.   It is also possible to achieve depletion in the 90-99% range using higher reagent concentrations (not shown).   Note that non-specific depletion remains unaffected over this broad range of FF. Since quantitation is done by examining labelled cells remaining in the supernatant and further as no wash of the wire loops was done in these studies, it seems from the constancy of the NSB values that they are inflated. This is consistent with the results obtained for data of Table II where the labelled cells (only) were included as controls.   Also note that the final dilutions of ferrofluid/anti-CD45 reagent were actually one-half the values indicated above due to the 1:1 dilution with cells in the final reaction mixtures.

The results of using another MAB (anti-CD4) ferrofluid reagent which also binds to receptors on this cell line and the effect of labelling reagent concentration on depletion are given in Figure 7. The reaction mixtures and conditions were identical to that of the previous study except that the amount of the GAMFc coated on this ferrofluid was substantially lower (50%) than that used in the anti-CD45 studies. As with anti-CD45 ferrofluid, depletion is a function of labelling reagent concentration.

Table II.   Effect of Cell Labelling Reaction Time on Specific and
Non-Specific Depletion of T-Cells

Labelling Time (minutes, rm. temp.)	non-specific [a]	% Depletion non-specific [b]	specific [c]
15	4.98	5.60	78.5
30	4.68	6.05	77.8

a.   Control with ferrofluid <u>buffer only</u> plus T-cells.

b.   Control with ferrofluid - <u>no</u> MAB plus T-cells.

c.   Ferrofluid/anti-CD45 reagent plus T-cells.

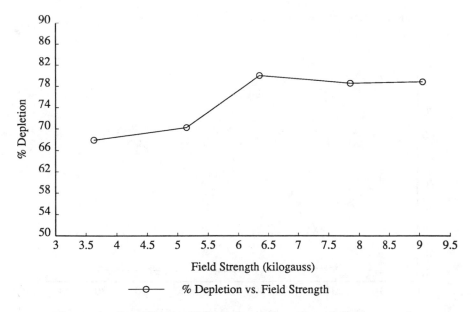

Figure 4.   Specific T-cell depletion as function of field strength.

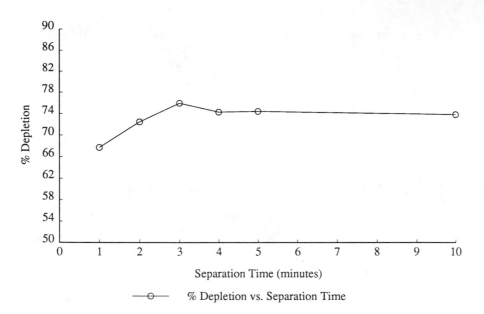

Figure 5.   Specific T-cell depletion as a function of high gradient magnetic separation time using an electromagnet at 7.0 kG.

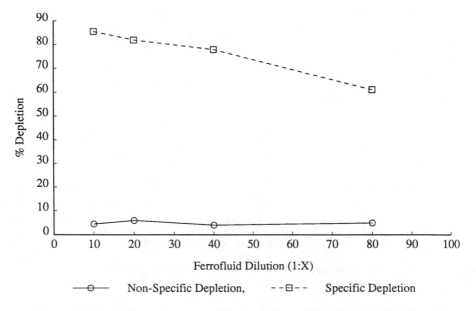

Figure 6.   Specific and non-specific depletion of T-cells as a function of ferrofluid concentration using anti-CD45 Mab.

To determine the potential of this system to remove a sub-population of cells from whole blood, preliminary experiments were done on mixtures of labelled T-cells and sheep red blood cells where the latter were added as a complicating factor. Microscopic examination of such mixtures showed no specific cell type interactions (rosetting). In these experiments T-cells were labelled with ferrofluid/anti-CD4 reagent either in the absence or presence of red cells. For example, two methods of introducing the ferrofluid/MAB reagent were used. In Method 1, the [51]chromium labelled T-cells were incubated with the ferrofluid/anti-CD4 reagent (reagent at 1:5 in CB3c) first for 30 minutes at $4^{o}C$ followed by the addition of the sheep red blood cells. In Method 2, [51]chromium labelled T-cells were combined with the sheep red blood cells followed by the addition of the ferrofluid/MAB reagent which was incubated for a 30 minute period as above. In both cases the ferrofluid volume was 125 ul (1:5) plus 100 ul sheep red blood cells (packed form in alsever's buffer) plus 25 ul [51]chromium labelled T-cells. Also in both cases, the high gradient magnetic separation was performed for 5 minutes using the wire loop system in a 7.5 kG permanent magnetic arrangement. These results are set forth in Table III which shows that the presence of even this high concentration of red cells in the labelling or separation step is without effect. Considering the level of red cells present these results are impressive.

Labelling and Depletion of T-Cell Subsets from Whole Blood Lysate and Whole Blood. The tests described below were performed to determine if ferrofluid and the wire loop magnetic separation system has some utility in immune status monitoring. Although the [51]chromium cell labelling scheme used in previous studies could have been applied here, it would not have provided specific data on the various cells in whole blood or whole blood lysate without adding significant complications to the analysis system. Therefore, fluorescent MAB's in conjunction with flow cytometry were used to analyze cell populations before and after depletions. The use of fluorescent MAB's and flow cytometry is commonly used to quantify cell populations and subpopulations in whole blood lysate or peripheral blood mononuclear cells. High gradient magnetic separations were done on cells from whole blood and whole blood lysate using ferrofluids specific for cell surface antigens (e.g. BioReceptor Ferrofluid + anti-CD45). In all cases, aliquots of the cells prior to any treatment were analyzed by flow cytometry to obtain values for lymphocytes, monocytes and granulocytes. By comparing these values with those obtained for cells remaining in the supernatants after high gradient separation, a determination of the % depletion of the various cell types was obtained.

For example, in an experiment to determine the percent of CD4 positive cells that were depleted from whole blood lysate, the value for the total amount of CD4 positive cells was first quantitated in an unseparated aliquot using a fluorescent anti-CD4 MAB and flow cytometry. Separation was performed using BioReceptor Ferrofluid with non-fluorescent anti-CD4 and the supernatant analyzed by staining cells with fluorescent anti-CD4 MAB and running the flow cytometer. Percent depletion would then be calculated by the difference between the original values and that found in the supernatants following magnetic separation. Care was taken to assure that the anti-CD4 (non-fluorescent) used with the ferrofluid for separation was not in excess but just below the particle capture capacity. This is stressed because if the "separating" anti-CD4 MAB was in excess it would bind CD4 positive sites on unseparated cells which would therefore not be available for the detection via

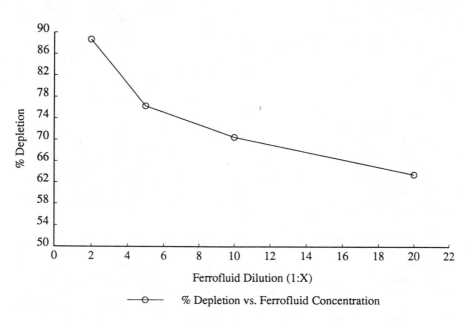

Figure 7.   Specific depletion of T-cells as a function of ferrofluid concentration using anti-CD4 Mab.

Table III. T-Cell Depletion From A Suspension Of Packed Sheep Red Blood Cells (SRBC's) Using Ferrofluid With Anti-CD4 MAB

Method [a]	Number of T-cells/test	Number of SRBC's/test	% Specific Depletion
1	2.5×10^5	2×10^9	80.7
2	2.5×10^5	2×10^9	80.1

a.   Method 1 where T-cells were incubated with the ferrofluid/anti-CD4 MAB reagent prior to combining with SRBC's; Method 2 where T-cells and SRBC's were incubated first followed by introducing the ferrofluid/anti-CD4 MAB reagent.

the subsequent fluorescent anti-CD4 MAB labelling. The whole blood lysate and whole blood used in these studies was obtained from healthy donors.

Table IV depicts the % depletion for separations done with control MAB (non-specific MAB), with Pan T-cell MAB (anti-CD45) and with helper cell MAB (anti-CD4). Lines a. and b. are results for a low and high level of GAMFcFF respectively using whole blood lysate and line c. for an experiment using whole blood and the high level of GAMFc FF. All reaction mixtures consisted of 125 ul of diluted ferrofluid containing either non-specific IgG, anti-CD45 or anti-CD4 plus 125 ul of either whole blood lysate or whole blood. Whole blood lysate was obtained via red blood cell lysis and white cells were washed 2 times in PBS (isotonic) and resuspended to 2×10^6 c/mL. Whole blood was used without any prior treatment (collected in EDTA vacutainers). At the low level of GAMFc, 74% of the bright staining T-helper cells were depleted while 33% of the dims were removed. At this GAMFcFF level 84% of T-cells were removed specifically as were 23% of the phagocytic monocytes. At the higher FF level specific depletion increased but so did monocyte depletion. For whole blood significant quantities of helper (58%) and T-cells (73%) were removed while for all the separations non-specific depletion in these complex mixtures was relatively low.

High Gradient Magnetic Separation With Radial Open Field Gradients From Quadrupole Electromagnet. The efficiency of the open field radial magnetic gradient generated by a quadrupole electromagnet (which previously served as a focusing component of a linear accelerator) was tested as a separation system for cells. This was accomplished by fitting the original 4, 12cm long pole pieces of the quadrupole with additional pole pieces to obtain an air gap of 1.6cm into which could be placed a 20mL cell culture tube.

For this experiment a 50nm dextran ferrofluid particle synthesized to have high non-specific cell binding (by modification of the Molday method) was used to demonstrate the feasibility of the separation system. The separation could be done <u>without</u> a specific MAB due to the high non-specific binding of this particle. These results, shown in Table V, demonstrate that even with this relatively low gradient, cells can be pulled to the sides of a 1.5cm i.d. tube and thus demonstrates the feasibility of using this system to separate cells. Although this experiment was performed with non-specific binding ferrofluid, from early work with the wire loop system it was found that such depletions correlate very nicely with the specific depletion one obtains using a low non-specific binding ferrofluid and specific MAB. Currently, permanent magnet quadrupoles and appropriate vessels for batch and large scale continuous separations are being developed. It should be noted that in this arrangement where labelled cells are pulled to the sides of a vessel, recovery of such cells is straightforward.

Summary

Methods for the synthesis of coated BioReceptor Ferrofluids have been developed. These reagents are superparamagnetic crystals of magnetite coated with protein of two kinds, a non-active protein such as serum albumin which serves only a coating function and a second protein or a bio-receptor such as MAB or a lectin which gives the reagent specificity. These coated ferrofluids show low non-specific binding to cells and other biological material. Further

Table IV.   Depletion of CD45 and CD4 Positive Cells From Whole
            Blood Lysate (WBL) and Whole Blood (WB)

% Non-Specific Depletion FF-Non-Specific IgG$_1$			% Specific Depletion FF-anti-CD45			FF-anti-CD4	
Lymphs	Monos	Grans	Lymphs	Monos	Grans	Brgts	Dims
a. 6.8	14.5	3.6	84.2	23.5	8.4	73.6	33.1
b. NS	2.0	NS	99.0	69.0	36.0	97.0	66.0
c. NS	8.3	NS	72.7	7.7	23.0	57.5	ND

a. Ferrofluid with a low level of GAMFc used with either non-
    specific IgG, anti-CD45 or anti-CD4 to deplete target cells
    from WBL.

b. As in a. except ferrofluid contained a high level of GAMFc.

c. As in b. except depletion performed from WB.

Key

Abbreviations	Description
Lymphs	Lymphocytes
Monos	Monocytes
Grans	Granulocytes
Brgts	Bright CD4 Positive Lymphocytes
Dims	Dim CD4 Positive Monocytes
ND	Not Done
NS	None Seen
FF	Ferrofluid

Table V.   Non-Specific Depletion as a Function of Field Strength and
Separation Time Using a Quadrupole Electromagnet

Field Strength (kiloguass) [a]	Separation Time (Minutes)	% Non-Specific Depletion [b]
5	5	35.5
5	10	61.8
5	15	71.1
6	5	58.3
6	10	75.6
6	15	86.9

a. Field strength varied by adjusting the voltage to the electromagnetic
Quadrupole system.

b. % depletion determined by removing a small aliquot from the 20mL
reaction vessel at 5, 10 and 15 minutes separation time, counting
the cells from these supernatants on a hemocytometer; a control tube
was set up with no ferrofluid and was sampled at 5, 10 and 15 minutes
to establish the control values; reaction mixtures consisted of 10mL of
dextran ferrofluid diluted 1:10 in PBS + 10mL of T-cells in CB3 at
5×10^5c/mL.

the coated ferrofluid magnetically labels target cells. In concert with two novel classes of high gradient magnetic separators [ferromagnetic rigid structures for generating internal gradients and quadrupoles for generating radial open field gradients] the potential of these systems in the affinity separation of cells has been demonstrated in a preliminary but thorough manner. Some of the systems examined and reported herein are relatively "clean" model cellular systems selected to optimize for some of the variables in the separation process while other systems reported are for more complex cellular matrices such as whole blood lysate and even whole blood. Being able to perform separations from both simple and complex mixtures greatly expands the applications to which these separation systems can be applied. It is hoped that these developments make available to the cellular experimentalist some new tools to add to the existing armamentarium of technologies for cell separation. In some cases these methods or modifications of them may suffice in providing a one-step purification. Because of the complexities of cellular systems, more than likely they will be used in concert with other well established or emerging methods described in this volume.

Literature Cited

1. Bean, C.P. and Livingston, J.D., 1959, *J. Appl. Phys.* 30, 1205.
2. Elmore, W.C., 1938, *Physical Review* 54, 1092.
3. Hirschbein, B.L., Brown, D.W. and Whitesides, G.M., 1982, *Chemtech* 1982 172.
4. Kaiser R. and Miskolczy, G., 1970, *J. Appl. Phys.* 41 106.
5. Kolm, H.H., 1975, *IEEE Trans Magn.* 11 1567.
6. Kronick, P.L., 1980, in *Methods in Cell Separation* (N. Catsimpoolas, Ed.) 3 115, Plenum Press, N.Y.
7. Liberti, P.A. and Pino, M.A., 1989, U.S. Patent Application No. 379,106.
8. Margel. S., Beitler, U. and Ofarim, M., 1982, *J. Cell Sci.* 56 157.
9. Mellville, D., Paul, F. and Roath, 1975, *Nature* 255 706.
10. Miltenyi, S., 1980, *J. Immunol. Meth.* in press.
11. Molday, R.S. and MacKenzie, D., 1982, *J. Immunol. Meth.* 52 353.
12. Molday, R.S. and Molday, L.L., 1984, *FEB's Lett.* 170 232.
13. Oder, R.R., 1976, *IEEE Trans Magn.* 12 428.
14. Owen, C.S. and Sykes, N.L., 1984, *J. Immunol. Meth.* 73 41.
15. Owen, C.S. and Liberti, P.A., 1987 in *Cell Separation: Methods and Selected Applications"* (Pretlow and Pretlow, eds.) 4 259, Academic Press, N.Y.
16. Owen, C.S., Silvia, J.C., D'Angelo, L.D. and Liberti, P.A., 1989, U.S. Patent No. 4,795,698.
17. Owen, C.S., 1978 *Biophys J.* 22 171.
18. Sharpe, P.T., 1988, *Methods in Cell Separation*, Elsevier, Oxford.
19. Treleaven, J., Gibson, F., Ugelstad, J., Rembaum, A. and Kemshead, J., 1984, *Lancet i*, 70.
20. Ugelstad, J. and Berge, A., 1988, *Fres. Zeit. Anal.Chem.* 330-4 328.
21. Whitehead, R.A., Chagnon, M.S., Groman, E.V. and Josephson, L., 1985 U.S. Patent No. 4,554,088.
22. Yau, J.A., Reading, C.L., Thomas, M.W., Davaraj, B.M., Tindle, S.E., Jagannath, S. and Dicke, K.A., 1990, *Exp. Hematol.*, 18 219.

RECEIVED March 15, 1991

INDEXES

Author Index

Affiliation Index

Subject Index

Production: Margaret J. Brown
Indexing: Deborah H. Steiner
Acquisition: Cheryl Shanks
Cover design: Lori Seskin–Newman

Printed and bound by Maple Press, York, PA